Marketing By Menu

Marketing By Menu
Third Edition

Nancy Loman Scanlon

JOHN WILEY & SONS, INC.
New York Chichester Weinheim Brisbane Singapore Toronto

This book is printed on acid-free paper. ⊗

Copyright © 1999 by Nancy Loman Scanlon. All rights reserved.

Published by John Wiley & Sons, Inc.

Published simultaneously in Canada.

This publication is designed to provide accurate and authoritative information in regard to the subject matter
covered. It is sold with the understanding that the publisher is not engaged in rendering professional services. If
professional advice or other expert assistance is required, the services of a competent professional person should
be sought

Library of Congress Cataloging-in-Publication Data
Scanlon, Nancy Loman.
    Marketing by menu / Nancy Loman Scanlon. — 3rd ed.
      p.   cm.
    Includes bibliographical references and index.
    ISBN 0-471-25330-8 (alk. paper)
    1. Menus. 2. Food service—Marketing. 3. Restaurants—Marketing.
  I. Title.
TX911.3.M45S38   1998
647.95′068′8—dc21   98-8155

Printed in the United States of America.

10   9   8   7   6   5   4   3

This edition of *Marketing By Menu* is dedicated to the memory of Barry Thompson, chef-owner of The Sail Loft in Rockport, Maine and former executive chef of the Washington Hilton. His willingness as an executive chef to teach the importance of a knowledge of kitchen operations in the activity of menu program development was a key factor in the career development of many of us who had the pleasure of working with him. For those who had the opportunity to come under his influence, he instilled a joy for good food and song.

Barry Thompson

1941–1997

◆ CONTENTS

◆ PREFACE

Since 1985, when the first edition of *Marketing By Menu* was published, the face of the restaurant industry in the United States and internationally has changed dramatically. The export of the American restaurant franchise, whether McDonalds or Planet Hollywood, has created international recognition of concepts and business practices originating from the United States. In 1998, the foodservice industry in the United States estimated sales of $336 billion ($U.S.) and supplied 10.2 million jobs to the workforce (National Restaurant Association). Restaurants today provide a social resource and cultural setting for the dramatic changes in lifestyle that have swept across the United States in the last quarter of the twentieth century. Communication and technology have helped to carry these same changes to a global audience and demands for the foodservice industry with them. This third edition of *Marketing By Menu* reflects many of those changes in the selection of menus of the contributing international restaurants and hotels featured in these pages.

Computer applications of business practices in the foodservice industry now supply daily and weekly reports and business summaries, and create up-to-the-minute inventory and cost analysis, providing foodservice operators the ability to create more profitable and successful operations. Because technology rapidly outdates itself, we have chosen not to include references to computer systems and software in this edition, featuring instead the reporting systems and concepts on which computer applications are founded.

This edition of *Marketing By Menu* has been designed for maximum use by foodservice professionals and educators alike. Work pages to facilitate an understanding of concepts such as menu repertory development and sales mix analysis have been included in this edition. New concepts and trends are recognized along with emerging markets in the foodservice industry, among them "home replacement foods," the expansion of catering services, the casual family-style restaurant explosion, and the importance of children as restaurant customers. *Marketing By Menu* III establishes the importance of the menu in creating profitable and successful restaurant and foodservice operations, detailing the many steps—from market survey to sales history—needed to achieve and maintain profitable operations.

ACKNOWLEDGMENTS

This third edition continues to owe a great deal to those individuals who have contributed to the overall body of knowledge covered in this text since its original publication. The gathering together of new menus and materials for this edition could not have been accomplished without the generous support of the restaurants, hotels, and foodservice operations represented in these pages. In particular I would like to thank Paul Wise, Chairman of the Hospitality, Restaurant, and Institutional Management program at the University of Delaware for his continued encouragement and support, and my colleagues, Ali Poorani, Bob Nelson, Pamela Cummings, Francis Kwanza, Ron Cole, Jim Meyers, Jim Lynch, and Joe DiGregorio. Many thanks to my fellow members of CHART (Council of Hotel and Restaurant Trainers) who have helped in the gathering of menus, and for the input and professional guidance of the hospitality education community. The assistance of Frank West of Ad Art Litho in Cleveland, Ohio is primary to my having been able to make the many changes in the menu graphics of this edition. Radisson Hotels were very generous in having menus submitted from their properties internationally, an effort that has allowed me to bring exciting graphics and design concepts to this edition.

As I travel, both in the United States and internationally, the willingness of foodservice operators to share their ideas through the pages of this and other industry reference books that I have authored makes my work a delight to accomplish on a daily basis. From San Diego, California to Camden, Maine . . . from Seattle, Washington, to Key Biscayne Florida . . . from Manila, Philippines to Stockholm, Sweden . . . from Cairo, Egypt to the villages of Brecon and Ut in Wales, the menus in these pages reveal the energy, entrepreneurship, creativity, culture, food ways and business practices of the foodservice community internationally. I am very grateful to them all for the contributions of their work to _Marketing By Menu_ III.

◆ INTRODUCTION

The objective of this third section of *Marketing By Menu* is to provide a thorough understanding of the importance of a well-developed menu program to the success of a foodservice operation. Marketing feasibility studies; food costing and pricing; the menu item selection process; and menu marketing and design techniques are just some of the major topic areas discussed in this edition. To help in understanding the food costing and menu development process, work pages provide the opportunity to practice some of the concepts and problems presented in the text.

The menu is the major selling tool of any restaurant. The attitude of management and owners is reflected in the quality of menu planning and design. How well the menu suits the community profile is reflected in management's ability to answer the needs of its customers. Menu pricing concepts, a flexible menu selection process, daily and seasonal menus, "healthy eating" menu programs, along with catering and take-out/delivery options are some of the trends in menu development and engineering discussed in the third edition of *Marketing By Menu*.

The foodservice professional in the twenty-first century is faced with the challenge of meeting the needs of the most diverse customer group that the American public has ever presented. As one of the largest industries in the United States, foodservice businesses produce a sizeable percentage of the gross national product and, in the franchise division, serve as one of the United States's largest exports to the international market. Our industry provides long-term employment to Americans from every walk of life, level of education, nationality, gender, and age. As America continues to shift its focus from manufacturing to service-related businesses, foodservice will continue to maintain a leading edge in revenue growth, employment, training, and development for the American work force.

To answer the needs of America's changing population, the foodservice industry is constantly developing new products and services. By the year 2000, 75 million Americans will be between the ages of 50 and 75. As the largest population segment in the United States, this group will continue to demand increased residential and health care facilities. Their interests are diverse and challenging, ranging from sports fitness and personal health to education and travel. Many will continue to be part of the American work force well beyond traditional retirement age, with needs centered on economically priced, high-quality, nutritious foods. Take-out and delivery services will be especially appealing to this age group.

As two-income and single-parent families have changed the profile of the working population, so has the role that the foodservice industry plays in the way America eats. This trend has resulted in a significant increase in the need for preprepared foods consumed outside the home or "taken-out to eat-in" as working families delegate the responsibility for food preparation to "home-food replacement providers." In the 1970s, the American work force was promised more leisure time in the

1990s. The opposite of this prediction has become the reality, with Americans now working longer hours in five- and often six-day weeks as business has reengineered labor into a smaller and more effective work force. To the foodservice industry, this means a greater demand for a wide variety of product. Families and individuals now need recreation opportunities within a two-hour distance from home that offer the possibility of two- and three-day mini-vacations. Hotels, resorts, and restaurants in both rural and urban communities provide a variety of foodservice opportunities to this customer base.

Foodservice industry operations include both commercial restaurants and institutional outlets. Commercial foodservice operations range in style from public restaurants to private-club dining and are segmented into the general categories of full service, casual style, fast food, theme, and specialty cuisine. Institutional foodservice must now meet the challenge of satisfying consumer demands, adapting the full range of restaurant techniques to settings as diverse as hospitals, retirement communities, school and colleges, industrial plants, and office buildings.

The restaurant of the twenty-first century will likely become a destination for socializing over quality foods, a way for the diverse American family to gather around a table that will also provide a source of entertainment. The foodservice professional of the year 2000 must be prepared to respond to new sources of food products, continuing developments in cuisine offerings, and the challenges of attracting labor and training to meet increasingly high customer expectations. In addition, new technology will be presented for food preparation, communication, and business operations. All of these elements must be considered and integrated into businesses while management strives to make a profit. Meeting these many needs and maximizing on new opportunities requires a knowledge of foodservice business practices. The application of technology to make foodservice businesses more accurate, timely, and profitable only serves to increase the challenge and drive the need for education and training in all aspects of the foodservice industry.

◆ PART I ◆

Profitable Menu Planning

CHAPTER 1

♦

The Commercial Menu

The elaborate, colorful, highly glossed menus offered in today's restaurants have a history that dates back only to the late 1800s, when Parisian restaurateurs introduced the carte, or bill of fare, to their patrons. (The carte à payer was the check, or bill, for the meal.) The menu for the day was drawn on a large poster-sized placard and placed conspicuously at the front door of the restaurant to draw in customers. French artists of the period, such as Renoir, Gauguin, and Toulouse-Lautrec, sketched and painted cartes. Toulouse-Lautrec is famous for his caricatures of Parisian women, some of whom were models for his cartes. In Figure 1-1, he uses a woman to illustrate a dinner menu. Renoir produced the novel carte shown in Figure 1-2, showing a chef juggling the day's menu, written on plates. In general, however, cartes were handwritten without much thought for appearance and were intended only to inform the customer of the offering for the day. Printing costs were prohibitive at that time. The practice of providing each guest with a menu did not become popular until the early 1900s.

The standard restaurant menu of today is based on the established format of course offerings for each meal. The classical dinner format consists of seven to eight course offerings with a variety of menu items (Figure 1-3). Figure 1-4 shows a prix fixe or

♦ **FIGURE 1-1.** Parisian artist Toulouse-Lautrec illustrated this dinner menu. (Hale 1968)

◆ **FIGURE 1-2.** In this menu, the artist Renoir depicts a chef juggling his menu for the day. (Hale 1968)

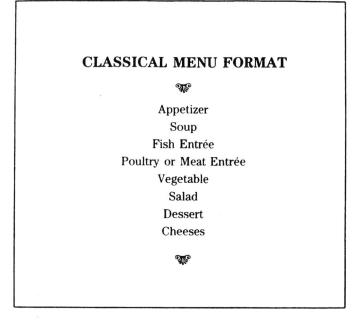

CLASSICAL MENU FORMAT

Appetizer
Soup
Fish Entrée
Poultry or Meat Entrée
Vegetable
Salad
Dessert
Cheeses

◆ **FIGURE 1-3.** The classical eight-course dinner format.

LLANGOED HALL
Suggested Four Course Dinner Menu
Sunday 12th January 1997

Terrine of Confit Guinea Fowl and Red Wine Shallots
with Pickled Root Vegetables

*****

Roast Fillet of Monkfish on Warm Potato Salad
with a Jacqueline Sauce

*****

Breast of Duck with Creamed Spinach, Celeriac
and a Port Wine Sauce

*****

Chocolate Tart with Crème Anglaise

*****

Coffee and Sweetmeats

£29.50

We respectfully request that should you wish to
smoke, you do so in the public rooms and
not in the dining room. Thank you.

♦ **FIGURE 1-4.** A fixed price or prix fixe menu for a four course dinner at Llangoed Hall, Brecon, Wales. Reprinted with permission.

fixed price menu, where one price is charged for the entire meal. Although the format has been modified by eliminating the salad and cheese courses, the general order of classical course presentation has been adhered to.

Different nations provide variations on this format with course addition based on the diet and cuisine of the country involved. The Italian menu includes a pasta course following the appetizer, while the French menu places an egg course between the fish and beef courses. In American cuisine, the egg was mostly limited to the breakfast table but emerged, in the 1970s, on brunch and supper menus. Fruits have gained prominence in the latter part of the French menu and appear only occasionally in the appetizer section of the American menu. Spanish and Oriental menus include rice as a course, while Middle Eastern menus often emphasize other grains, such as couscous. Russian and Scandinavian menus include a large smorgasbord for the appetizer course, with many different kinds of meats, fish, cheeses, breads, spreads, and vegetable preparations in small portions with a variety of beverages. This variety of cuisine and availability of international foods make modern meal planning a creative and exciting field.

A current variation of the classical menu format is shown in Figure 1-5. On this menu, titles are provided for each course and the classical menu format is adapted to suit French cuisine. As in classical French menu terminology, the appetizer course is "Les Hors d'Oeuvre Varies," or a variety of hors d'oeuvre. The appetizer is followed by a small portion of fish, a clear soup, a light veal dish, sorbet, and the main entree of filet of beef with accompaniments. Salad and an assortment of cheeses finish the meal with dessert following. Wines have been listed for each course. A contemporary variation of this format used for the menu of Outback Steakhouse can be seen in Figure 1-6. Outback Steakhouses offers a selection of items for each course, presented in the

La Reception

Les Hors D'Oeuvres Varies

Le Diner

Prologue
Le Pave de Saumon aux Caviar et Concombres

Ouverture
Le Consomme Double aux Palmiers

Divertissement
La Ris de Veau Paloise

Interlude
Le Sorbet au Melon

Piece de Resistance
Le Filet de Boeuf Fleuriste

Les Pommes Chateau   Les Tomates Jardini

Pastorale
Le Salade de Saison

Pas de Cinq
Les Fromages de Provinces .

Finale
Les Oeufs Chocolat aux Nids de Sucre

Epilogue
Les Mignardises
La Demi-Tasse

♦ **FIGURE 1-5.**   The classical menu format as adapted for a commercial banquet menu offering French cuisine.

◆ FIGURE 1-6. A variation on the classical format is presented in this menu from the Outback Steakhouse. Reprinted with permission of Outback Steakhouse.

order in which they are served. The menu from the Green Room, the Hotel du Pont, Wilmington, Delaware, in Figure 1-7 outlines four of the classical courses for a la carte service. Vegetables are included with the entree and desserts are displayed on a rolling cart for guests to choose from.

Commercial foodservice operations should structure menu formats around the needs of the customer as well as the limitations of the kitchen. Figure 1-8 illustrates the simplest menu format for restaurant operations. Variations on this format will depend on cuisine, theme, or specialization of the restaurant and the courses offered on the menu.

The luncheon menu format shown in Figure 1-9 is an abbreviated dinner format, which can be further adapted to include interesting course variations to make the menu more appealing and reflect the dietary needs of midday customers. Recently, drastic changes have occurred in the eating habits of the business person, who typically eats a main meal in the evening and has become more conscious of dietary needs, both health and weight related. The menu in Figure 1-10 addresses these needs, reflecting management's awareness of consumer needs. Panel one of this menu explains the res-

Brandywine Room Menu

—To Start—

Shrimp Mousse Crab Cake, Champagne Mustard Sauce
9.00

Grilled Portabella Mushroom, Asiago Cheese,
Basil Oil, Sundried Tomato Pesto
7.50

Butternut Squash Risotto with Pancetta, Porcini Mushrooms,
Pumpkin Seed Oil
8.00

Smoked Salmon Tartare, Lemon Emulsion and Rye Toast Points
9.00

Maryland Crab Bisque
6.00

Mesclun Salad with Passion Fruit Vinaigrette,
Walnuts and Gorgonzola
7.50

Traditional Caesar's Salad
per person, 9.00

—Bottled Water—

Perrier, 25 oz.
4.00

Poland NonCarbonated, 28 oz.
4.00

A 17% service charge will be added to parties of 7 or more.
Thank you. To ensure the comfort of other guests, we request that you refrain from smoking. Thank You

—Entrees—

MEAT

Grilled Filet Mignon, Mushroom Ragout, Cabernet Sauce
29.00

Roast Rack of Domestic Lamb, Minted Vegetable Couscous, Pan Jus
29.00

Honey Roasted Free Range Chicken with Thyme Jus
20.00

Seared Breast of Pekin Duck, Port Wine Peppercorn Sauce
24.00

FISH & SEAFOOD

Crab "Norfolk"
28.00

Sesame Crusted Tuna with Lemon Grass Broth
24.00

Baked Salmon with Horseradish Crust, Creamed Leeks
and Celery Root Purée
21.00

Lobster Thermidor Molly Brown
30.00

Chilean Sea Bass with Lemon-Basil Mashed Potatoes,
Dried Tomato and Olive Tapenade
23.00

VEGETARIAN DU JOUR

Tortellini Provençale
20.00

~ Ask our Staff about Tonight's Chef Specials ~

♦ **FIGURE 1-7.**   The commercial presentation of four classical courses for a la carte service. From the Brandywine Room of the Hotel duPont, Wilmington, Delaware. Reprinted with permission of the Hotel duPont, Wilmington, Delaware.

taurant's food preparation philosophy and healthy eating program. Panel two identifies 'Healthy Heart' menu items with a heart beside the item name.

American cuisine, with its "Great American Sandwich," promotes a simplified luncheon format of soup, sandwich, and dessert. This format is specifically geared toward offering fast service of three courses in a limited dining time. The menu in Figure 1-11 is a daily, single-panel menu following the luncheon menu format offering salads and soups in the appetizer section, then sandwiches, meal-sized salads, and reduced portion sizes of entrees. The dessert menu is offered on a separate card. Both menus are produced on a color laser printer. The breakfast menu flows in and out of historical record, appearing without much consistency. Certainly, everyone

DINNER MENU FORMAT

Appetizer

Soup

Salad

Entrées
Fish
Poultry
Meat

Vegetable
Starches

Dessert

◆ **FIGURE 1-8.** A simplified dinner format, popular with commercial foodservice operations.

LUNCHEON MENU FORMAT

Appetizer

Soup

Salad

Eggs

Entrées
Fish
Poultry
Meat

Vegetables
Dessert

Cheese

◆ **FIGURE 1-9.** Simplified luncheon format, with cheese and egg courses offered in place of starches.

♦ **FIGURE 1-10.** A restaurant menu featuring a healthy eating program. From the Silver Diner in Rockville, Maryland. Reprinted with permission of the Silver Diner Restaurant.

has always "broken the fast" at some time during the morning hours, but the degree of prominence given the first meal of the day has depended on culture and nationality.

The simplest breakfast of fruit, roll, and coffee or tea is known as a continental or European breakfast. The Yankee who, in the 1800s, started the day with "black tea and toast, scrambled eggs, fresh spring shad, wild pigeon, pig's feet, two robins eggs

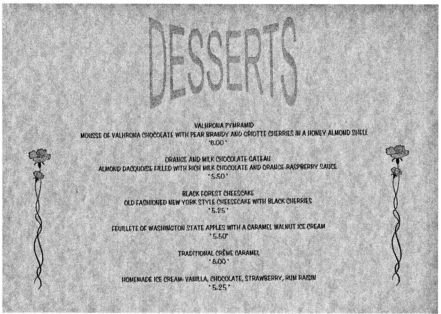

◆ **FIGURE 1-11.** The luncheon menu format shown on the menu from the Green Room at the Hotel duPont, in Wilmington, Delaware. Reprinted with permission of the Hotel duPont, Wilmington, Delaware.

on toast, and oysters" clearly needed something a bit more substantial. A breakfast menu from an early-nineteenth-century cookbook by Mistress Margaret Dods is even heartier and follows the classical presentation of fish, meats, and sweetbreads (Figure 1-12)

The typical hotel breakfast menu from 1920 shown in Figure 1-13 is both complete

Nineteenth-Century American Breakfast

Oatmeal with Cream

Smoked Herring
Sardines with Mustard
Broiled Trout

Cold Meat Pies
Broiled Kidneys
Scotch Woodcock

Sausage with Mashed Potatoes
Tongue with Horseradish Sauce

Singing Hinnies
Bannocks
Barmbrack
Honey, Conserves, Marmalade
Tea   Coffee
Gentlemen only: Ale, Rum, and Scalch

♦ **FIGURE 1-12.**   A breakfast menu from the early-nineteenth-century cookbook by Mistress Mary Dods follows the classical presentation of fish, meats, and sweetbreads. (Singing hinnies are a kind of griddlecake; bannocks are flat cakes made from cornmeal; and barmback is bread or cakes made with currants.) (Hale 1968)

Apples   Malaga grapes   Oranges
Radishes
Clam broth
Cracked wheat or boiled rice
Eggs to order
Omelet with asparagus tips
Broiled brook trout, Montpelier butter

BROILED
Tenderloin steak with mushrooms
Lamb kidneys with bacon
Quail with watercress
Sausage   Fried oysters

POTATOES
Stewed in cream   French-fried
Baked
Vienna rolls   Toast   Cornbread
Cream biscuits   Buckwheat griddle cakes
Preserved strawberries
Coffee   Tea   Cocoa

OR
Cantaloupe
Oatmeal
Sliced cucumbers
Broiled trout   Salt mackerel
Broiled tenderloin or sirloin steak
Fried spring chicken, cream sauce
Ham   Bacon
Boston baked beans, brown bread
Eggs to order
Omelets, plain or with ham

POTATOES
Baked   Stewed in cream
French-fried
Wheat cakes
Parker House rolls   Horn rolls
Corn muffins
Coffee   Tea   Chocolate

♦ **FIGURE 1-13.**   In comparison with this hotel breakfast menu from 1920, breakfast in the 1990s is a very simple meal.

and well designed. By comparison, breakfast in the 1990s is a very simple affair. The contemporary breakfast menu may concentrate on standard items or develop brunch items that can be served later in the morning (Figure 1-14).

Breakfast menus offer menu planners an opportunity to develop creative breakfast packages that feature local specialties. A New England menu might offer a "Yankee breakfast" of eggs, bacon, baked beans, muffins, apple pie, and coffee. In Savannah, Georgia, a southern-style breakfast might include sliced ham, beaten biscuits, and grits.

Figure 1-15 is the breakfast menu of a typical American family-style restaurant and offers breakfast combination meals in a semi a la carte format with beverages, fruits, and cereal priced a la carte. Figure 1-16, a hotel breakfast menu, upgrades both presentation and prices but follows the same format. American breakfast menus do not usually include fish or vegetables.

Many full-service restaurants that once limited their meal services to luncheon and dinner are expanding to include breakfast on a seven-day, five-day or weekend schedule. Increased revenue from breakfast service requires only the added expense of labor, utilities and food cost. This source of revenue can supply a high profit margin if the turnover rate is sufficient to generate a significant cash flow. Good breakfast revenues can help offset low customer counts and average checks during other meal periods. Breakfast has become an increasingly popular meal for the restaurant industry and rates careful consideration in menu development.

♦ **FIGURE 1-14.** The contemporary breakfast menu may include standard breakfast fare or more elaborate brunch items.

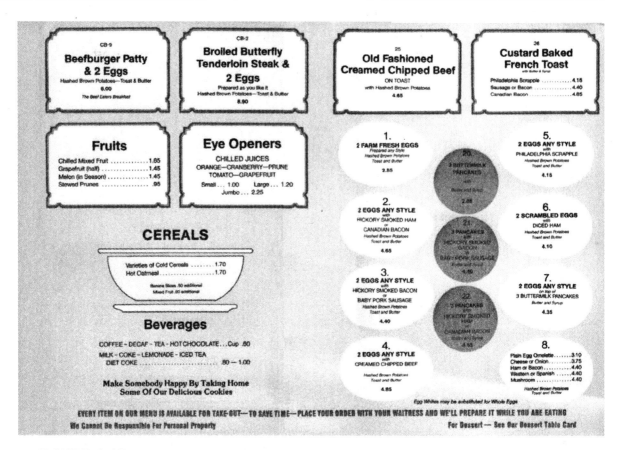

◆ **FIGURE 1-15.** A typical American family-style restaurant breakfast menu from the Melrose Diner in Philadelphia, Pennsylvania. Reprinted with permission of Melrose Diner, Philadelphia, Pennsylvania.

THE DAILY MENU VERSUS THE DAILY SPECIAL

The standard restaurant menu of the 1990s is very different from its 1960s counterpart. The same can be said for the profile of the American restaurant consumer. Demanding variety, value for dollar, entertainment, and quality, customers are putting increased pressure on foodservice operators at all levels to compete with new and better menu items and promotions.

In the 1980s, a trend developed toward offering the full-service restaurant customer a menu designed on a daily basis, for which items are chosen according to seasonal availability, customer preference, and profitability. This is in contrast to the standard practice of offering daily specials that are either posted on a menu board, attached to the menu, or recited by the server. In the 1990s, this trend has grown, encouraged by the availability of software design packages and high-quality laser printers on-premise.

The Terrace Dining Room

BREAKFAST

Full Buffet Includes all items from the Buffet Table and your choice of one Breakfast Specialty, Coffee, Tea and Juice	18.75
Buffet Includes all items from the Buffet Table and your choice of Coffee, Tea and Juice	14.75

Starters

House-Smoked Salmon Herbed Cream Cheese and Toasted Bagel	12.25	O Fresh Fruit, Seasonal Berries Date-Nut Bread	8.25
O Parma Ham and Cantaloupe	8.75	Warm Oatmeal, Raisins and Toasted Almonds	5.25
O Pink Grapefruit and Strawberries	5.75	O Tropical Fruit Smoothie	5.50
Cream of Wheat with Berries	6.25	O Strawberry-Banana Smoothie	5.50

Waffles and Pancakes

Housemade Cheese Blintzes with Strawberries	8.50	Belgian Waffles Strawberries and Devonshire Cream	9.25
Sun-Dried Cherry French Toast with Pine Nuts Vermont Maple Syrup	9.50	O 7-Grain Pancakes Wild Berries and Mountain Honey	8.95
Blueberry Pancakes Vermont Maple Syrup	8.95	Banana and Pecan Malted Waffles Pecan Maple Butter	9.25
Buttermilk Pancakes	8.00		

Breakfast Specialties

Poached Eggs with Corned Beef Hash and Marinated Tomatoes	10.50
Phoenician Huevos Rancheros	10.00
Southwestern Egg and Potato Skillet	10.25
Eggs Benedict with Hollandaise Sauce and Hash Browns	11.75
Three Egg Omelette with Choice of Fillings	10.00
Artichoke and Spinach Frittata	10.25
O Vegetarian Egg White Frittata	10.25
O Egg White Omelette with Tomato Salsa and Herbed Goat Cheese	9.75
Two Eggs, Any Style Breakfast Potatoes and Choice of Toast	8.25

All frittatas and omelettes can be prepared with egg whites, for your well-being.

On The Side

Housemade Spiced Sausage	5.75	One Egg, Any Style	3.75
Hickory Smoked Bacon	5.75	Hash Browns	3.75
Grilled Country Ham	5.75	Choice of Toast	2.50
Corned Beef Hash	5.75	Bagel with Cream Cheese	4.00

Arizona birds are affectionate, friendly and bold. May we suggest that you do not feed them.

As a courtesy to other diners, please refrain from using cellular telephones in the restaurant.

O Choices – Superb cuisine created by Chef Jeffrey Gosselin, with your well-being in mind.

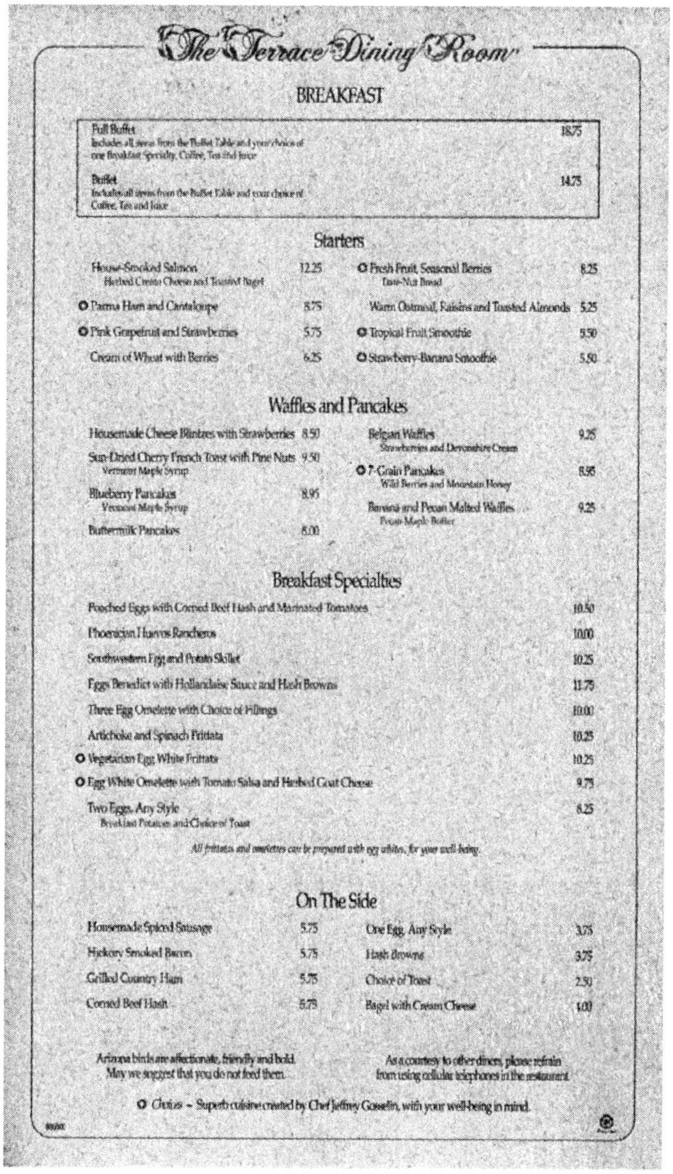

♦ **FIGURE 1-16.** A hotel breakfast menu from The Phoenician Hotel in Scottsdale, Arizona offering fish, meat, egg and vegetable breakfast items. Reprinted with permission of Phoenician Hotel, Scottsdale, Arizona.

Figure 1-17 is a menu featuring daily food and wine specials. The chef and restaurant manager review the selections weekly for product availability and determine which items will be put into production. Figure 1-18 also has a daily menu format featuring seafood. Of the eleven entrees in the "daily" section, nine are seafood. Figure 1-19 uses the same printing process for a daily menu using a well-designed, preprinted paper. Figure 1-20 uses a single-panel menu printed in-house,

♦ **FIGURE 1-17.** A single-panel menu that includes a daily selection of food and wine. From the Georgetown Seafood Grill, Washington, D.C. Reprinted with permission of Georgetown Seafood Grill, Washington, D.C.

♦ **FIGURE 1-18.** The menu from Etta's Restaurant in Seattle, Washington features a daily menu section as a supplement to the on-going menu. Reprinted with permission of Etta's Restaurant, Seattle, Washington.

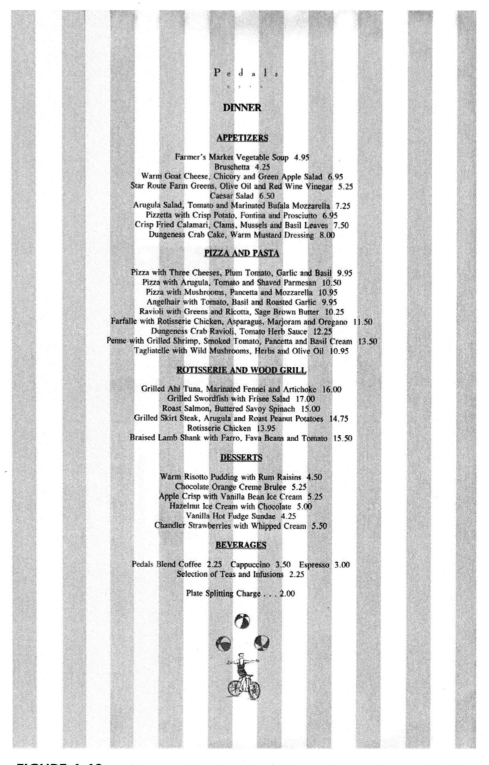

P e d a l s

DINNER

APPETIZERS

Farmer's Market Vegetable Soup 4.95
Bruschetta 4.25
Warm Goat Cheese, Chicory and Green Apple Salad 6.95
Star Route Farm Greens, Olive Oil and Red Wine Vinegar 5.25
Caesar Salad 6.50
Arugula Salad, Tomato and Marinated Bufala Mozzarella 7.25
Pizzetta with Crisp Potato, Fontina and Prosciutto 6.95
Crisp Fried Calamari, Clams, Mussels and Basil Leaves 7.50
Dungeness Crab Cake, Warm Mustard Dressing 8.00

PIZZA AND PASTA

Pizza with Three Cheeses, Plum Tomato, Garlic and Basil 9.95
Pizza with Arugula, Tomato and Shaved Parmesan 10.50
Pizza with Mushrooms, Pancetta and Mozzarella 10.95
Angelhair with Tomato, Basil and Roasted Garlic 9.95
Ravioli with Greens and Ricotta, Sage Brown Butter 10.25
Farfalle with Rotisserie Chicken, Asparagus, Marjoram and Oregano 11.50
Dungeness Crab Ravioli, Tomato Herb Sauce 12.25
Penne with Grilled Shrimp, Smoked Tomato, Pancetta and Basil Cream 13.50
Tagliatelle with Wild Mushrooms, Herbs and Olive Oil 10.95

ROTISSERIE AND WOOD GRILL

Grilled Ahi Tuna, Marinated Fennel and Artichoke 16.00
Grilled Swordfish with Frisee Salad 17.00
Roast Salmon, Buttered Savoy Spinach 15.00
Grilled Skirt Steak, Arugula and Roast Peanut Potatoes 14.75
Rotisserie Chicken 13.95
Braised Lamb Shank with Farro, Fava Beans and Tomato 15.50

DESSERTS

Warm Risotto Pudding with Rum Raisins 4.50
Chocolate Orange Creme Brulee 5.25
Apple Crisp with Vanilla Bean Ice Cream 5.25
Hazelnut Ice Cream with Chocolate 5.00
Vanilla Hot Fudge Sundae 4.25
Chandler Strawberries with Whipped Cream 5.50

BEVERAGES

Pedals Blend Coffee 2.25   Cappuccino 3.50   Espresso 3.00
Selection of Teas and Infusions 2.25

Plate Splitting Charge . . . 2.00

◆ **FIGURE 1-19.**   This menu is printed on a laser printer using preprinted menu stock. From Pedals Restaurant in the Radisson Hotel, LaJolla, CA. Reprinted with permission of Radisson Hotels International.

BURGERS AND SANDWICHES

We grind choice chuck roast in-house daily for our burgers.
Served with a choice of iron skillet beans, French fries, cole slaw or couscous

CHEESEBURGER..Loaded old fashion style with mustard, mayonnaise, lettuce, tomato, pickle and onion....... 10
HICKORY BURGER..Canadian bacon, onion, homemade hickory sauce and cheddar cheese.................... 10
TEXAS BURGER..Our Firehouse chili, with cheddar cheese and chopped onion. Saturday only! 10
VEGGIE BURGER..Amazing vegetarian recipe with oat bran and brown rice on a toasted bun 10
CLUB SANDWICH..Ham, chicken, cheddar and Jack cheese, bacon, lettuce, tomato and mayonnaise 11
BILTMORE SPECIAL..Grilled chicken breast, red onion, Jack cheese, mayonnaise on a toasted bun 11

SALADS AND VEGETARIAN PLATTER

Our own signature salad dressings: Mustard Honey, Blue Cheese, Buttermilk Garlic, and Classic Vinaigrette.

TRADITIONAL SALAD.. Crisp greens, chopped egg, tomatoes, homemade croutons and bacon.................... 8
SOUP and SALAD.. Today's signature soup, with our Traditional or Caesar salad.................................. 10
GRILLED CHICKEN SALAD.."The Original" with honey lime vinaigrette and homemade peanut sauce........ 12
THE CLUB SALAD..Crisp greens, lightly fried chicken, bacon, chopped egg, tomatoes and avocado 11
VEGETARIAN PLATTER..Seasonal fresh vegetables, brown rice and black beans.................................... 12
SEARED TUNA SALAD..Seared rare with cilantro ginger vinaigrette, field greens and mango 13
EGGLESS CAESAR.. Chopped romaine, croutons, our signature dressing and Reggiano parmesan............... 8

> ### TODAY'S FISH
>
> We apply the highest standards to guarantee the freshest fish available.
> All seafood is carefully inspected at the time of arrival to ensure absolute quality.
> Market Price
>
> SEARED AHI TUNA STEAK..With Asian vinaigrette and couscous
> TODAY'S FRESH FISH..Fileted in-house, with cole slaw or couscous
> FRESH GRILLED SALMON..With cole slaw or couscous

ENTREES

ROASTED CHICKEN WITH COUSCOUS..One half chicken oven-roasted, basted with natural juices .. 13
NEW YORK'S BEST PORK CHOP..Finished over hard wood with mashed potatoes and salad 21
BARBECUE RIBS.."Our knife and fork version." Danish pork ribs with French fries and cole slaw 22
ROAST PRIME RIB.. Aged prime rib with baked potato and salad. Horseradish sauce upon request........ 22
HOUSTON'S HAWAIIAN.. Marinated ribeye grilled over hardwood with baked potato and salad 22
NEW YORK STRIP..Center cut, prime beef, hardwood grilled with baked potato and salad...................... 23
STEAK ST. CHARLES..Sliced grilled prime tenderloin, celery mashed potatoes, and vegetable broth 19
CHICKEN TENDERS PLATTER..Lightly fried chicken tenderloins with French fries and cole slaw.............. 14
BARBECUE CHICKEN.. Backyard style! Grilled boneless breast with French fries and cole slaw.............. 13
ROASTED CHICKEN WITH BLACK BEANS..One half chicken with melted Monterey Jack cheese........ 14

THIS AND THAT

"HOME SMOKED" SALMON..With fresh toasts and our Chef's dressing.....9
"CHICAGO STYLE" SPINACH DIP....................................9
JUMBO SHRIMP COCKTAIL....................14

SIGNATURE SOUPS made daily...................5	CHICKEN TENDERS BASKET.....................8
COLE SLAW..3	LOADED BAKED POTATO.........................4
IRON SKILLET BEANS................................3	SEASONAL FRESH VEGETABLES..............MKT
FRENCH FRIES..3	COUSCOUS...3
MASHED POTATOES..................................3	BLACK BEANS WITH BROWN RICE............3

DESSERTS

All desserts $6

HOT FUDGE SUNDAE..Homemade and very tasty
KEY LIME PIE..with fresh whipped cream
FIVE NUT BROWNIE..a la mode with Kahlua
APPLE WALNUT COBBLER..with French vanilla ice cream

ESPRESSO.......2   CAPPUCCINO.......3

♦ **FIGURE 1-20.** The menu from Houston's Restaurant in New York offers daily fish specials based on market availability. Reprinted with permission of Houston's Restaurant, New York.

but does not produce it on a daily basis. The "Today's Fish" section in the center of the menu can be changed as needed to reflect market availability.

How a restaurant chooses to promote its menu program depends primarily on:

♦ customer demand
♦ product availability

♦ the flexibility of the kitchen staff
♦ management's ability to plan profitable "special" menu items

A daily menu planned on a cycle basis to repeat itself over a predetermined time period will help to cut down on the amount of time necessary to design these menus cost effectively. No matter how creative, exciting, or innovative menu programs are, they must accomplish the most important goal of any restaurant—customer satisfaction.

CYCLE MENUS

Commercial foodservice organizations usually base their menu format and the structure of their operations on a menu that remains essentially the same day after day, until rising costs or customer demands necessitate a new menu. Some restaurants take pride in their ability to offer the same menu year after year. If the customer audience changes often enough, such as in tourist locations, or if repeat customers can find enough variety in the basic menu to be satisfied, this system can work. The manager and the chef may introduce daily or weekly specials and change a few soups and desserts to provide a bit of variety. In a situation, however, where the customer base is made up of consistent repeaters and menu items must be limited for cost and/or service reasons, cycle menu programs can provide variety to an otherwise static menu.

A cycle menu is an organized schedule for presenting preplanned menus in a repeated pattern over a number of days. The format was designed principally for institutional or cafeteria foodservice organizations, but it is often successfully incorporated into commercial menu programs.

Each restaurant should review its specific menu needs and design a total menu format accordingly. Incorporating the cycle menu format into a commercial menu can provide additional cost controls while simultaneously introducing variety for both staff and customers. Since the cycle menu format is essentially a mode of scheduling, the same rules of menu planning apply to cycle menus as to all other types of menu formats.

The basic initiative for adopting a cycle menu is to provide a varied, balanced, and nutritional diet. Many foodservice institutions serve a "captive audience"— consumers who cannot leave to seek alternative dining and must take whatever is offered. If these people were given the same food on a daily or regular basis, they would become bored, restless, and finally angry. To alleviate this problem, the cycle menu is tailored to offer variety and change in meals. How much variety is provided depends on such considerations as budget, purchasing capability, storage, preparation, and food availability. Four different types of menu cycles are commonly used— typical, typical-break, random, and split.

Typical

What makes a cycle menu typical is that the same menu is always served on a given day of the week. If the captive audience changes, as in a resort hotel or a hospital, this pattern can be continued over a long period of time with variations imposed only for dietary or seasonal accommodations.

Typical-Break

A typical-break cycle differs from the typical cycle in that a limited number of menus are presented in a pattern that repeats itself on staggered days of the week. A day school, for instance, that offers a four- or five-day meal plan could adopt a six-day menu cycle, repeating the menus in various ways. Figure 1-21 illustrates a six-day menu cycle served over a four-day meal plan. As the pattern continues into the third block, a typical menu structure appears, repeating every 13 days.

Random

The random cycle presents an interesting variation of menus for the extended captive audience. These could include students, military personnel, prison inmates, extended-stay hospital patients, or rest-home clients. The random cycle includes a complete day's menu for every day of the period chosen, from as few as 10 to as many as 365. The menu planners then assign letters or numbers to each menu and schedule the sequence in which menus will be served, basing the decision on season, availability of foods, and popularity of menus. If turkeys were suddenly to become available in bulk at good wholesale prices, for example, the menu planner would slot in the menus that could take advantage of the purchase. This schedule format may seem like the typical cycle at first, but in practice offers diversification with preplanned and precosted menus for each meal service. Figure 1-22 shows an example of the random menu is use for a seven-day meal plan with a 26-day cycle. This cycle can be rotated by changing the letters around. Problems with format arise when menus that contain similar items are placed next to each other. The program must be monitored to ensure that each menu has been reviewed to determine its effect on the next one.

TYPICAL-BREAK CYCLE

	MONDAY	TUESDAY	WEDNESDAY	THURSDAY
WEEK 1	DAY 1	DAY 2	DAY 3	DAY 4
WEEK 2	MONDAY DAY 5	TUESDAY DAY 6	WEDNESDAY DAY 1	THURSDAY DAY 2
WEEK 3	MONDAY DAY 3	TUESDAY DAY 4	WEDNESDAY DAY 5	THURSDAY DAY 6
WEEK 4	MONDAY DAY 1	TUESDAY DAY 2	WEDNESDAY DAY 3	THURSDAY DAY 4
WEEK 5	MONDAY DAY 5	TUESDAY DAY 6		

◆ **FIGURE 1-21.** In this typical-break cycle, a six-day cycle of menus is presented over a four-day meal plan.

RANDOM-CYCLE MENU						
DAY 1	DAY 2	DAY 3	DAY 4	DAY 5	DAY 6	DAY 7
A	J	L	Z	C	D	M
DAY 8	DAY 9	DAY 10	DAY 11	DAY 12	DAY 13	DAY 14
P	I	B	Q	N	E	W
DAY 15	DAY 16	DAY 17	DAY 18	DAY 19	DAY 20	DAY 21
F	O	H	R	V	K	S
DAY 22	DAY 23	DAY 24	DAY 25	DAY 26	DAY 27	DAY 28
U	G	X	T	Y	A	J

◆ **FIGURE 1-22.** Each letter in this random-cycle menu schedule represents a different menu. The cycle can be changed by rearranging the order of the letters in the schedule.

Split

The split cycle offers even greater possibilities of variety, but with far less control. In a split-cycle menu, each item of the menu is given a cycle of its own, to be used in conjunction with the other items. Items are placed in a cycle according to their popularity with the consumer group; the menu planner responds directly to a group's likes and dislikes when scheduling. Figure 1-23 shows a group of items from a hospital's split-cycle menu.

SPLIT-CYCLE MENU		
		DAY 1
S	15	Cream of Spinach Soup
E	12	Roast Pork
V	7	Buttered Beets
P	5	Baked Idaho Potato
D	10	Angel Food Cake
		DAY 2
S	6	Chicken Vermicelli
E	9	Swiss Steak
V	4	Green Beans
P	6	Boiled Parsley Potato
D	12	Rice Pudding
		DAY 3
S	5	Minestrone Soup
E	7	Spaghetti with Meatballs
V	3	Green Salad
D	5	Chocolate Sundae

Letters identify food category: soup, entrée, vegetable, dessert. Numbers indicate item rotation.

◆ **FIGURE 1-23.** In this split-cycle menu, each item has a cycle of its own. Reprinted with permission of Rhode Island Hospital.

In the split-cycle format, the consumers see the items that they prefer more often, in this case baked potatoes, green beans, minestrone soup, green salad, and chocolate sundaes. Roast pork and rice pudding, being less popular, are served every 12 days. The number of days for the entire cycle is determined by the general restrictions of the foodservice organization.

Although the split-cycle format offers unlimited menu planning possibilities, it poses problems in creating appetizing balanced menus, since items in one cycle may be incompatible with other menus scheduled for the same day. Constant review is needed. This menu form also restricts the ability of the menu planner to precost a menu completely unless the following steps can be taken:

1. *The combinations* are carried out ahead of time.
2. *Menu items* in each category are priced within a given range so that all combinations in the menu category will cost out acceptably.
3. *Management* will find food-cost fluctuations acceptable, realizing that the food-cost percentage will level off to predetermined goals at the end of the period.

AIRLINE MENUS

Airlines use cycle menus as their primary menu form. In many cases, only one menu plan is used on a major route—specific regional cuisines on long-term or international flights, for example, or more general fare for carriers shuttling between metropolitan areas of the country. The rate at which the customer repeats a travel route determines the need for variety in menu plans. In order to accommodate the frequent business flyer as well as the occasional traveler with specific dietary needs, airlines have instituted meal reservation plans. When a customer makes a flight reservation, a meal request reservation can be made at the same time. American Airlines offers a total of 22 to 23 menu selections. The breakdown of offerings is as follows:

1. *Dietary meals:* Ten selections
   Low calorie
   Low cholesterol
   Low carbohydrate
   Low sodium
   Vegetarian
   Bland/soft
   Diabetic
   Kosher
   Hindu
   Moslem
2. *American travelers' meals:* Six selections
   Cold seafood platter
   Fresh fruit bowl
   California quiche
   Fresh vegetable plate

Great American hamburger
Honey-dipped chicken
3. *Children's menu:* Four items
Hotdogs
Hamburgers
Peanut butter and jelly
Fried chicken
4. *Regular menu:* Two to three items depending on availability

Frequent business travelers have the opportunity to make meal requests from this selection in order to vary their diet when they frequent the same travel route. Passengers with specific dietary needs can also avail themselves of a full selection of menu items. Along with these choices, a menu is offered that will repeat itself on a daily basis for a specified period of time. Airlines work with individual catering companies in each city, supplying specifications, recipes, cost requirements, and quality control inspection.

COMMERCIAL RESTAURANT MENUS

Commercial restaurant menus can adopt cycle menus formats for several different purposes:

♦ Special entree selections (daily or weekly basis)
♦ Dessert menus (weekly, ten-day, or split cycle)
♦ Vegetable and starch menus (weekly, 10-day, or split cycle)

A good example of the use of a combination of cycle formats is shown in the ten-day buffet menu cycle in Figure 1-24. The Soup selection is identified by a ten-day cycle, while the Salad selection remains the same on a daily basis, with only the content of the medley changing on an availability basis. The Specialties section remains the same on a daily basis and entrees rotate on a ten-day cycle. Starches have a three-day cycle with potato on day 1, rice on day 2, and orzo on day 3, and Vegetables vary daily according to freshness and availability. The Carver station always offers roast turkey. The Dessert section offerings are occasionally supplemented by specialties from the pastry kitchen, but generally remain constant.

As seen by these ten daily menus in Figure 1-24, any part of the commercial restaurant menu can be adapted to the cycle schedule. The primary menu should refer to the separate dessert menu or daily specials section, if they are in use.

Lunch Buffet # 1

Soup
Amish Chicken and Corn

Salads
Cous cous with Baby Peas
Marinated Mushrooms with Leeks
Medley of Greens with Mushroom, Red Pepper and House Vinaigrette

Specialties
Sugar Maple House Smoked Trout with Capers and Onions
Fresh Herb Oven Roasted Tomatoes
Assorted Cheeses with Fresh Fruit
Roasted Garlic Tapanade with Toasted Baguette
Northern White Bean and Tomato Salad

Entrees
Pecan Crusted Pork Loin with Sundried Pears and Braised Cabbage
Penne Pasta with Roma Tomatoes, Black Olives, Mushrooms, Onions, Garlic and White Wine.
Oven Roasted Potatoes
Fresh Vegetable Medley

Assorted Breads and Rolls with Cinnamon Honey Butter and Whipped Butter

Carved Specialty
Home Smoked Turkey Breast with Whole Grain and Dijon Mustard, Fresh Herb Mayonnaise, and Cranberry Relish

Desserts
Orange Creme Brule
Assorted Mini Pastries
Chocolate Torte
Assorted Cookies
Amaretto Bread Pudding

Lunch Buffet # 2

Soup
Vegetable Barley

Salads
Cous cous with Baby Peas
Marinated Mushrooms with Leeks
Medley of Greens with Dried Pears, Toasted Almonds and House Vinaigrette

Specialties
Sugar Maple House Smoked Trout with Capers and Onions
Fresh Herb Oven Roasted Tomatoes
Assorted Cheeses with Fresh Fruit
Roasted Garlic Tapanade with Toasted Baguette
Northern White Bean and Tomato Salad

Entrees
Breast of Chicken with Sundried Cranberries and Toasted Pinenuts
Cheese Tortellini with Mushrooms, Tomatoes and Cream
Oven Roasted Potatoes
Fresh Vegetable Medley

Assorted Breads and Rolls with Cinnamon Honey Butter and Whipped Butter

Carved Specialty
Home Smoked Turkey Breast with Whole Grain and Dijon Mustard, Fresh Herb Mayonnaise, and Cranberry Relish

Desserts
Orange Creme Brule
Assorted Mini Pastries
Chocolate Torte
Assorted Cookies
Amaretto Bread Pudding

Lunch Buffet # 3

Soup
Minestrone

Salads
Cous cous with Baby Peas
Marinated Mushrooms with Leeks
Medley of Greens with Bacon Twists, Feta Cheese and House Vinaigrette

Specialties
Sugar Maple House Smoked Trout with Capers and Onions
Fresh Herb Oven Roasted Tomatoes
Assorted Cheeses with Fresh Fruit
Roasted Garlic Tapanade with Toasted Baguette
Northern White Bean and Tomato Salad

Entrees
Breast of Chicken with Artichoke Hearts, Plum Tomatoes, White Wine and Scallions
Seschwan Beef Stirfry
Basmati Rice Pilaf
Fresh Vegetable Medley

Assorted Breads and Rolls with Cinnamon Honey Butter and Whipped Butter

Carved Specialty
Home Smoked Turkey Breast with Whole Grain and Dijon Mustard, Fresh Herb Mayonnaise, and Cranberry Relish

Desserts
Orange Creme Brule
Assorted Mini Pastries
Chocolate Torte
Assorted Cookies
Amaretto Bread Pudding

Lunch Buffet # 6

Soup
Cheddar Leek

Salads
Cous cous with Baby Peas
Marinated Mushrooms with Leeks
Medley of Greens with Mandarin Oranges, Caramelized Walnuts and House Vinaigrette

Specialties
Sugar Maple House Smoked Trout with Capers and Onions
Fresh Herb Oven Roasted Tomatoes
Assorted Cheeses with Fresh Fruit
Roasted Garlic Tapanade with Toasted Baguette
Northern White Bean and Tomato Salad

Entrees
Breast of Chicken with Andouille Sausage, Sundried Tomatoes and Cream
Flank Steak with Braised Cabbage
Oven Roasted Potatoes
Fresh Vegetable Medley

Assorted Breads and Rolls with Cinnamon Honey Butter and Whipped Butter

Carved Specialty
Home Smoked Turkey Breast with Whole Grain and Dijon Mustard, Fresh Herb Mayonnaise, and Cranberry Relish

Desserts
Orange Creme Brule
Assorted Mini Pastries
Chocolate Torte
Assorted Cookies
Amaretto Bread Pudding

Lunch Buffet # 7

Soup
Chicken and Rice

Salads
Cous cous with Baby Peas
Marinated Mushrooms with Leeks
Caesar Salad with Homemade Croutons

Specialties
Sugar Maple House Smoked Trout with Capers and Onions
Fresh Herb Oven Roasted Tomatoes
Assorted Cheeses with Fresh Fruit
Roasted Garlic Tapanade with Toasted Baguette
Northern White Bean and Tomato Salad

Entrees
Breast of Chicken Veracruz with Black Olives, Tomato, Cumin and Jalapenos
Penne Carbonara
Rice Pilaf with Toasted Almonds
Fresh Vegetable Medley

Assorted Breads and Rolls with Cinnamon Honey Butter and Whipped Butter

Carved Specialty
Home Smoked Turkey Breast with Whole Grain and Dijon Mustard, Fresh Herb Mayonnaise, and Cranberry Relish

Desserts
Orange Creme Brule
Assorted Mini Pastries
Chocolate Torte
Assorted Cookies
Amaretto Bread Pudding

Lunch Buffet # 8

Soup
Vegetable Orzo

Salads
Cous cous with Baby Peas
Marinated Mushrooms with Leeks
Medley of Greens with Shaved Romano, Red Pepper and House Vinaigrette

Specialties
Sugar Maple House Smoked Trout with Capers and Onions
Fresh Herb Oven Roasted Tomatoes
Assorted Cheeses with Fresh Fruit
Roasted Garlic Tapanade with Toasted Baguette
Northern White Bean and Tomato Salad

Entrees
Breast of Chicken Marbella with Spanish Olives and White Wine
Bowtie Pasta with Spinach, Sun Dried Tomatoes, Plum Tomatoes and Garlic
Rice Pilaf
Fresh Vegetable Medley

Assorted Breads and Rolls with Cinnamon Honey Butter and Whipped Butter

Carved Specialty
Home Smoked Turkey Breast with Whole Grain and Dijon Mustard, Fresh Herb Mayonnaise, and Cranberry Relish

Desserts
Orange Creme Brule
Assorted Mini Pastries
Chocolate Torte
Assorted Cookies
Amaretto Bread Pudding

◆ **FIGURE 1-24.** A commercial application of the cycle menu is seen in this 10-day buffet menu from Vita Nova restaurant at the University of Delaware. Reprinted with permission of the University of Delaware.

Lunch Buffet # 4

Soup

Cauliflower and Cavatelli

Salads

Cous cous with Baby Peas
Marinated Mushrooms with Leeks
Caesar Salad with Homemade Croutons

Specialties

Sugar Maple House Smoked Trout with Capers and Onions
Fresh Herb Oven Roasted Tomatoes
Assorted Cheeses with Fresh Fruit
Roasted Garlic Tapanade with Toasted Baguette
Northern White Bean and Tomato Salad

Entrees

Oven Roast Turkey with Toasted Pinenut Stuffing
Bowtie Pasta with Spinach, Sun Dried Tomatoes, Plum Tomatoes, and Garlic
Roasted Potatoes
Fresh Vegetable Medley

Assorted Breads and Rolls with Cinnamon Honey Butter and Whipped Butter

Carved Specialty

Home Smoked Turkey Breast with Whole Grain and Dijon Mustard, Fresh Herb Mayonnaise, and Cranberry Relish

Desserts

Orange Creme Brule
Assorted Mini Pastries
Chocolate Torte
Assorted Cookies
Amaretto Bread Pudding

Lunch Buffet # 5

Soup

Peppered Spinach and Clam Soup

Salads

Cous cous with Baby Peas
Marinated Mushrooms with Leeks
Caesar Salad with Homemade Croutons

Specialties

Sugar Maple House Smoked Trout with Capers and Onions
Fresh Herb Oven Roasted Tomatoes
Assorted Cheeses with Fresh Fruit
Roasted Garlic Tapanade with Toasted Baguette
Northern White Bean and Tomato Salad

Entrees

Grilled Breast of Chicken with Fresh Rosemary Salsa
Baked Haddock with Capers and White Wine
Roasted Potatoes
Fresh Vegetable Medley

Assorted Breads and Rolls with Cinnamon Honey Butter and Whipped Butter

Carved Specialty

Home Smoked Turkey Breast with Whole Grain and Dijon Mustard, Fresh Herb Mayonnaise, and Cranberry Relish

Desserts

Orange Creme Brule
Assorted Mini Pastries
Chocolate Torte
Assorted Cookies
Amaretto Bread Pudding

Lunch Buffet # 9

Soup

New England Clam Chowder

Salads

Cous cous with Baby Peas
Marinated Mushrooms with Leeks
Caesar Salad with Homemade Croutons

Specialties

Sugar Maple House Smoked Trout with Capers and Onions
Fresh Herb Oven Roasted Tomatoes
Assorted Cheeses with Fresh Fruit
Roasted Garlic Tapanade with Toasted Baguette
Northern White Bean and Tomato Salad

Entrees

Breast of Chicken with Wild Mushrooms, Madeira and Cream
Lasagna Roulade with Spinach and Goat Cheese
Oven Roasted Potatoes
Fresh Vegetable Medley

Assorted Breads and Rolls with Cinnamon Honey Butter and Whipped Butter

Carved Specialty

Home Smoked Turkey Breast with Whole Grain and Dijon Mustard, Fresh Herb Mayonnaise, and Cranberry Relish

Desserts

Orange Creme Brule
Assorted Mini Pastries
Chocolate Torte
Assorted Cookies
Amaretto Bread Pudding

Lunch Buffet # 10

Soup

Cream of Tomato Basil with Homemade Croutons

Salads

Cous cous with Baby Peas
Marinated Mushrooms with Leeks
Caesar Salad with Homemade Croutons

Specialties

Sugar Maple House Smoked Trout with Capers and Onions
Fresh Herb Oven Roasted Tomatoes
Assorted Cheeses with Fresh Fruit
Roasted Garlic Tapanade with Toasted Baguette
Northern White Bean and Tomato Salad

Entrees

Breast of Chicken with Parmesan Crust and Plum Tomato Sauce
Poached Salmon with Lemon Caper Vin Blanc
Wild Rice Pilaf
Fresh Vegetable Medley

Assorted Breads and Rolls with Cinnamon Honey Butter and Whipped Butter

Carved Specialty

Home Smoked Turkey Breast with Whole Grain and Dijon Mustard, Fresh Herb Mayonnaise, and Cranberry Relish

Desserts

Orange Creme Brule
Assorted Mini Pastries
Chocolate Torte
Assorted Cookies
Amaretto Bread Pudding

♦ **FIGURE 1-24.** *(Continued)*

CHAPTER 2

Planning
for
Profit

PREPARING THE MARKET SURVEY

The market survey, as discussed here, examines the major considerations for opening any foodservice operation. Almost every aspect of the survey that is considered for the success of a foodservice operation also applies to menus. The menu must be developed or revised according to customer and community demands, whether you are planning a new enterprise or changing an existing operation.

In using the market survey, each manager must decide which considerations apply to the individual foodservice outlet. No two situations are alike—resort and seasonal establishments need as thorough an investigation as year-round operations. Success in today's highly competitive and growing foodservice industry depends on ability, knowledge, and imagination. The market survey provides essential information; when properly used, it can help avoid time-consuming, costly pitfalls and aid in determining future expansion programs.

The major marketing tool of a foodservice operation is the menu, and the main function of the menu is to sell. The menu will sell as well as it is planned. Successful planning depends upon how well the contents of the menu—recipe cards, cost cards, and balanced menu items—are developed. An impressive, well-designed physical menu is important, but if it does not reflect the requirements of the operation or the purchasing power and established needs of the patrons, that menu will not be as salable and profitable as it should be.

Before any type of foodservice operation is established, a market survey should be done to determine three points: area need and ability to support the operation; target market; and patron desires or preferences. The survey evaluates three aspects of the market—the community, the customer, and the competition.

The Community

The first segment of the market survey, the community, looks at a general geographical area with two major questions in mind:

1. Is the location a concentrated urban or widespread suburban area?
2. What is the total population count for the geographical area, and where is the majority of the population concentrated?

The answers to these two questions tell prospective investors whether the location will provide patrons with easy access and a short travel distance, and whether there is a large enough concentration of the population within an established distance to support the operation. Location and transportation are major factors to be considered before opening a foodservice business. To establish population and travel statistics, define the location as closely as possible. Establish a radius, or distance, around the area in which survey results will be based. Determine the radius by using travel time as a guide. The average customer is willing to spend a half-hour traveling to a restaurant. Calculate this travel time using different types of transportation available to the customer and establish the radius around your location. What mode of transportation will the customer be using? If the foodservice outlet will be located in a downtown urban area, public transportation facilities must be considered.

Subway stations, bus stops, and taxicab availability are important. Customers who use public transportation are easier to measure because their travel radius is smaller. Location, in this situation, must be judged on its desirability: how willing are customers to travel into the area?

When the customer uses private transportation, accessibility of the outlet to highways, traffic arteries, and major intersections is a primary concern. Parking is an important consideration: are spots available on the street or is there a need for private lot accommodation? If parking is not readily available and a private lot must be used, who will assume the charges? If the location is far from major intersections, will that location make the travel effort worthwhile? Complicated parking and travel arrangements can seriously affect a customer's willingness to choose one restaurant over another for either private dining or banquet functions.

The economic status of a community is another important consideration. What are the major businesses in the area and who are the largest employers? What are the products being produced and what are the foreseeable markets for those products? What is the employment rate and the history of unemployment for the area? These answers can be used to forecast the future economic health of the area. If the community has been economically unstable over a number of years, it can be expected to become stable in the near future. To determine economic stability, the pattern of the community's economic growth over a long period of time must be reviewed. An instance of poor economic stability can be seen in the following review of one community's economic situation. Over the past 150 years, this city has seen many growth spurts, but due to a number of economic reversals has never had a continued period of well-being. The city is located directly outside a major metropolitan city, and at the turn of the twentieth century had a booming mill industry. When manufacturing costs moved mills elsewhere and foreign imports diverted the market, the industry faltered and then slowly died. Following World War II, major investments were made in developing the aerospace industry. Employment was up and the economy was growing; however, reversals in the 1960s and 1970s gradually forced this industry to begin slowing down production and reducing employment. Energy and manufacturing costs have now required plants to move to more economically feasible areas. Though still active in the area, the aerospace industry has been reduced drastically from the original postwar boom days. Although directly accessible to a major metropolitan city and located on the ocean, this city has never developed any associated tourist, fishing, or shipping industry. Stores and shops have moved to large malls to be more accessible to shoppers. Suburban housing development for major metropolitan workers has taken place around, but not in, the city, leaving a depressed urban community for customers. Unemployment figures are high and prospects for future industry growth questionable.

The economic history of this community is not a healthy one. Its seaside location might warrant consideration, but patronage would have to come from surrounding communities. This does not necessarily mean that this community could not support a successful restaurant. What it does mean is that an in-depth investigation into location and customer market must be made before a decision is reached.

Most communities publish a retail sales breakdown that shows the placement of consumer dollars in the categories of retail sales available in the area. A look at total food and beverage dollars spent in relation to total retail sales and total spendable income is a valuable key to the spending habits of a community and the willingness of that community to support restaurants.

The Customer

Knowing your customer beforehand will make it possible for the menu and restaurant to satisfy customer needs and desires to the fullest extent practicable. The vital statistics of a community form a picture of the average prospective customer. Census reports provide general population counts, age breakdowns, family household sizes, average incomes, expendable incomes, and other information that can help in determining the target market. A target market is the area or group of people that has been determined to be the primary customers. This category may be based on age, income, social or ethnic group, or a combination of these characteristics, and it is on this market that all of the selling efforts will be focused. Marketing is selling, and selling is done by responding to customer needs with the menu items, prices, atmosphere, and location this group will prefer.

The availability of a community's vital statistics depends upon the local chamber of commerce and business development groups. Communities and cities actively seeking new industries and businesses will have detailed reports available. In many areas, however, few, if any, statistics and population analyses are provided. In these cases, the U.S. Census would be the best source of information. Unfortunately, the national census is taken only once every ten years. Some communities periodically update the census figures, but others rely upon the last national census and use a percentage rate of growth or decline to calculate current figures.

The racial and ethnic breakdown of the community is important in determining national or religious cuisine preferences. If, for instance, the area has a large Italian population, there is most likely a market for Italian cuisine, especially if an evaluation of the competition shows that there are no full-service Italian cuisine restaurants. Although this is a good indication of a possible need, population statistics must be reviewed to determine the following points.

1. What is the average overall income and the average expendable income breakdown of this section of the population?
2. Will there be enough members of the overall population with a high enough expendable income to support this type of operation? Expendable income is the portion of a person's income that remains to be spent after necessities have been paid for. Necessities include rents or mortgages, utilities, food, education, transportation, basic clothing, taxes, and other living costs. If population statistics confirm that there is a population segment with a high level of expendable income, then the basis for developing the restaurant, as proposed, has been found. Investigation should also be made to see if any past efforts to establish this type of operation have taken place, and to see if these efforts succeeded or failed. If there were failures, ascertain whether they were due to poor management and insufficient foodservice knowledge, or lack of an adequate customer base. There is rarely only one reason for failure. Failures should be analyzed as thoroughly as successes, to make sure that the same mistakes are not repeated.

3. Finally, knowing the educational level of potential customers may help to determine their degree of food sophistication and acceptance. In general, the higher the level of education, the higher the degree of sophistication. Figure 2-1 offers a worksheet for researching your customer base.

The Competition

"Know your competition" is an important business precept. The competition must be established at the outset. The target area should be completely surveyed to determine the following:

The total number of foodservice operations in the area
A breakdown of the categories and numbers of foodservices available

---

MARKET AREA DEMOGRAPHICS

Use this worksheet to determine who your customer is.
The information that you need can usually be obtained from your local library, city hall, or chamber of commerce.

1. Area population growth over last 5 years:

2. Estimated population growth/decline in next 5 years:

3. Number of households:

4. Household income:

$75,000 or more	_____%
$50,000 to $75,000	_____%
$35,000 to $50,000	_____%
$25,000 to $35,000	_____%
$15,000 to $25,000	_____%
$ 7,500 to $15,000	_____%
Below $7,500	_____%

5. Average household size:

6. Per capita income:

7. Population by age   Marital status

18-25	_____%M	_____%S
25-35	_____%M	_____%S
35-40	_____%M	_____%S
40-50	_____%M	_____%S
50-65	_____%M	_____%S
65-70	_____%M	_____%S
70-80	_____%M	_____%S

Evaluate the results of your research to determine if there are enough potential frequent customers in your area to support your level of dining and the average check of your restaurant.

◆ **FIGURE 2-1.** Identifying target market area demographics is the objective of this worksheet. Reprinted with permission. Copyright 1991 Menu for Profit.

The general pricing structure of the menu for each operation

The established competition—those operations with the same pricing structure (average check) as your own

Once the competition has been determined, it must be examined in detail for its good and bad points. The competition survey sheet is a suggested formula for competition analyses (Figure 2-2). In answering these questions for every foodservice outlet that is considered competition, an overall picture of each competitor will develop. This will allow you to determine whether the needs of your target market are being met. If, for example, five restaurants are surveyed and all five serve primarily seafood, a new restaurant emphasizing meat and poultry might be well received. Additions such as entertainment, outside dining, a full-service lunch, or a well-developed wine and beverage list could give you the edge on the competition. Providing the missing elements are keys to a successful restaurant.

COMPETITION SURVEY SHEET

_Restaurant or Foodservice Outlet_____

_Address_____

_Date_____

_Completed by_____

1. LOCATION
   A. _Where in the community is the restaurant located?_____
   B. _Access from major roads_____
   C. _Parking_____
   D. _Sign Visible_____lighted_____
   E. _Access from public transportation_____
   F. _Availability of public transportation_____
   G. _Location features and liabilities_____

2. PHYSICAL APPEARANCE
   A. _Architectural style_____
   B. _Outstanding features_____
   C. _General exterior condition_____
   D. _General interior condition_____
   E. _Types of dining areas_____

3. SERVICE FEATURES
   A. _Days open (weekly)_____(yearly)_____
   B. _Hours open (daily)_____(weekly)_____
   C. _Seating capacity_____
   D. _Turnover (per dining area and meal service)_____
   E. _Average check (per meal)_____(overall)_____
   F. _General cuisine_____
   G. _Meals provided (breakfast)_____(lunch)_____(dinner)_____
      _(brunch)_____(late supper)_____(other)_____
   H. _Beverage services (lounge)_____(beer and wine only)_____
      _(bar)_____(full beverage list)_____
      _(wine list)_____(attached)_____
   I. _Service (good)_____(mediocre)_____(poor)_____
   J. _Additional facilities_____
   K. _Entertainment_____
   L. _Community acceptance_____
   M. _Apparent problems_____
   N. _Menu attached (yes)_____(no)_____
   O. _Floor layout (provide description or sketch)_____

◆ **FIGURE 2-2.** The competition survey sheet is a suggested format for analyzing competing foodservice operations in an area.

Attempting to duplicate a competitor's successful operation is not a sound foodservice practice. Although one success can be genuine and reflect the needs of the community and customer, too much of a good thing can saturate the market, dilute the novelty, and result in customer figures and profits that are lower than expected.

INTERPRETING THE MARKET SURVEY

The next important step in the market survey is the interpretation of its results in relation to your proposed foodservice plans. A review of the primary points covered in the market survey provides a feasibility analysis of the market.

1. *Community*
   a. What is the population size?
   b. What is the proposed location?
   c. Is that location accessible to major roads or public transportation?
   d. What is the overall economic health of the area?
   e. What portion of area total retail sales can be attributed to food and beverage service sales?
   f. What is the customer radius?
   g. Do enough local residents fall within that radius?
2. *Customer*
   a. What is the average age?
   b. What is the average income? Average expendable income?
   c. What is the breakdown of ethnic and religious population?
   d. What is the general educational level of the target market?
   e. What are the sources of possible customers from community businesses and organizations?
3. *Competition*
   a. How many foodservice operations are located in the area?
   b. What is the established competition?
   c. What does the target market lack?

After reviewing the market survey, compile a list of demands that a new restaurant must meet, based on the original proposal and the existing competition. You must now determine whether the available foodservice staff will be capable of producing the necessary cuisine and meeting the service requirements of your menu. Labor is one of the most significant problems facing the restaurant industry today. Will the local community be able to provide not only the quality but the quantity of both preparation and service staff that will be needed to run the restaurant for its proposed operating schedule? If, as an investor, you cannot meet the standards needed to compete successfully with existing foodservice operations, you should consider another area or community.

THE FEASIBILITY CHART

One of the biggest problems confronting potential restaurant and foodservice operators is establishing the general pricing structure that will be needed to make a profit.

(Doing good business is not necessarily the same thing as making a profit. A profit is the balance of funds that remains after all fixed and variable costs have been paid.)

Although institutional foodservices are nonprofit by definition, they usually return a small percentage of their revenue back to the operation to cover maintenance, repairs, equipment purchases, and expansion. This percentage is not defined as profit, however, because it is channeled back into the institution and because no individuals receive any of these funds. Commercial foodservices have no such restrictions on profit. There are, however, at least two limitations on the amount of profit that can be attained: the costs of goods and service, and how much the customer will pay. Often, some menu items generate more profit than others. With a properly developed menu and merchandising plan, the difference in profit potential can be balanced so that large profits on some items cover small profits on others. A properly developed and maintained food cost percentage helps generate the right amount of profit.

Large chain and franchise companies are founded on the idea of selling the same product in the same way to different groups of customers. The menu items and the general pricing structure are pre-established. If the market survey shows that the area will successfully support the restaurant, a decision is made to open. The menu prices reflect the current prices of the company and their direct competition; often, these chains are within pennies of each other on similar menu items.

An individual planning a foodservice operation can have a problem determining the menu prices needed to make a profit. Knowing what type of restaurant is being planned helps to establish the general pricing range. Although the market survey shows the prices that the competition charges for items similar to yours, these prices satisfy the needs of your operation. To charge the right price to ensure the proper profit, the following factors must be known:

1. Total amount of money needed (estimated costs) to run the operation for one year
2. The number of meal services (breakfast, lunch, and/or dinner) that will be available daily
3. The anticipated turnover rate for each one of the services
4. The number of seats in the restaurant
5. The number of days open annually
6. The number of customers to be served on an average day

A feasibility chart is a guideline that uses these six factors and the figures that they represent to establish the following financial factors for the foodservice establishment: (1) the average check, (2) the total amount of profit that can be generated, and (3) whether the foodservice establishment is feasible, or can make the amount of profit needed.

A feasibility chart is accurate only if all possible factors that could influence the profit margin are included. A sample feasibility chart is shown in Figure 2-4.

Estimated costs include all overhead, labor, food, and beverage costs. Overhead costs are expenses other than labor, food, and beverage costs (prime costs). The following are the categories of assigned overhead costs:

1. Direct operating expenses: expenses directly connected to providing service to the customer, including
   a. Linen, china, glass, and flatware
   b. Cost of cleaning and associated supplies
   c. Employee uniforms and laundry costs
   d. Cooking equipment, utensils, and kitchen tools
   e. Restaurant decorations, including flowers
   f. Menus, wine lists, and other promotional material
   g. Licenses and permits
2. Utility expenses: heat, light, and power, including any type of energy use (gas, oil, electricity and other canned or bottled fuel used in the restaurant). Water, ice and trash removal are also considered utility expenses
3. Administrative expenses: costs directly related to the operation of the business management of the restaurant, including
   a. Telephone services and equipment rental
   b. Computer and data processing costs
   c. Office supplies and printed materials used in the business management of the operation
   d. Insurance costs not related to either employee benefits or building occupational costs
   e. Management fees, such as accountants and property management
   f. Security services
   g. Professional dues and expenses
4. Advertising expenses: promotional costs, including
   a. Advertising through any form of media (newspapers, magazines, radio, television, or billboards)
   b. Cost of advertising or public relations agencies
5. Repairs and maintenance expenses: costs involved with the physical maintenance of the restaurant property, including
   a. Interior or exterior design costs
   b. Maintenance of furniture, upholstery, and window hangings
   c. Repairs to furnishings, floor coverings, and building structures
   d. Repairs to utilities such as refrigeration, air conditioning, plumbing, heating, kitchen equipment, and dishwishing and sanitation equipment
6. Occupation expenses (often called fixed costs): expenses that do not change regardless of the volume of the restaurant's business, including
   a. Rent (or mortgage)
   b. Property taxes
   c. Other taxes
   d. Property insurance (including fire, theft, and casualty)

Prime costs are the expenses involved with labor and the purchase of food and beverages. The following are assigned prime costs:

1. Food cost: direct costs related to the purchase of any food-related product used in the restaurant
2. Beverage cost: direct costs related to the purchase of any beverage-related product, alcoholic or nonalcoholic, used in the restaurant

3. Labor cost
   a. Wages and salaries including vacation pay, overtime, commissions, and bonuses paid to employees
   b. Costs of the following employee benefits:
      Social Security
      Federal and state unemployment taxes
      State health insurance tax
      Workmen's compensation insurance premiums
      Welfare payment plans
      Pension plans
      Insurance for health, accident, hospitalization; Blue Cross, Blue Shield, and other group policy premiums
      Expenses directly related to the well-being of the employee such as meals, transportation, education, and activities

The worksheet in Figure 2-3 is a statement of income and expenses that, when properly filled out with accurate sales and expense figures, can determine restaurant profit. This worksheet allows for the dollar value to be transferred to a percentage figure. Outlined are both prime costs and overhead costs as well as revenues from sales and interest. This worksheet should be used to estimate the total operating costs (combined overhead and prime) of the operation for the year. This total can then be used in the feasibility chart shown in Figure 2-4 to help accurately determine the probability of success of a proposed restaurant operation.

Ideally, profits should be calculated as 20 percent of total operating costs. This figure should also represent 10 percent of the total sales if this goal is to be met. In figure 2-4, the total costs are estimated at $1 million with a 20 percent profit margin of $200,000 and a total sales figure of $2 million.

The number of turnovers refers to the number of times that a seat will be occupied (i.e., "turned over" to a new customer) during a meal service. (A turnover of 1.5 for a dinner service is an average foodservice industry figure for a full-service restaurant.) The number of seats refers to chairs, not tables. The number of customers served during a day is calculated by adding the turnover rates for lunch and dinner and multiplying that figure by the total number of seats. This figure is the total number of customers served if the restaurant is filled as anticipated by the turnover rate. To make this figure more realistic, however, slow business periods must be accounted for and an average percentage of occupancy determined. The percentage of occupancy is a calculation of what percent of the total estimated customer base will actually be served by the restaurant.

The sample feasibility chart estimates that the percentage of occupancy should be reduced by 25 percent, making the occupancy rate 75 percent. The amount of reduction is based on the market survey and an analysis of the days of the year when business will be heavy or slow. Here the adjusted number of customers is 750 per day.

To find the adjusted total estimated number of customers to be served annually, multiply the average number of customers served daily by the number of days per year that the restaurant is open. To find the amount of money, or average check, that each customer must spend in order to cover the total costs of the operation,

WORSHEET
Complete and Compare
Statement of Income and Expenses

	Your Figures (Dollars)	Your Figures (% of Sales)	Restaurant Industry Report	Variance + or −
Sales				
Food	$	%	%	
Beverage	$	%	%	
Total sales	$	%	%	
Cost of sales				
Food	$	%	%	
Beverages	$	%	%	
Total cost of sales	$	%	%	
Gross profit	$	%	%	
Other income	$	%	%	
Total income	$	%	%	
Controllable expenses				
Payroll	$	%	%	
Employee benefits	$	%	%	
Direct operating expenses	$	%	%	
Music and entertainment	$	%	%	
Advertising and promotion	$	%	%	
Utilities	$	%	%	
Administrative and general	$	%	%	
Repairs and maintenance	$	%	%	
Total controllable expenses	$	%	%	
Income before occupation costs	$	%	%	
Rent	$	%	%	
Property Taxes	$	%	%	
Other Taxes	$	%	%	
Property Insurance	$	%	%	
Income before interest and depreciation	$	%	%	
Interest	$	%	%	
Depreciation	$	%	%	
Restaurant Profit	$	%	%	

Supplemental Operating Information	Your Figures	Restaurant Report	Variance + or −
Sales per seat			
Food	$	$	
Beverage	$	$	
Total	$	$	
Average receipt per cover (total sales ÷ covers)	$	$	
Daily seat turnover (covers ÷ seats ÷ 365 days)	times	times	
Index of productivity [sales ÷ (payroll + benefits)]	times	times	

♦ **FIGURE 2-3.** This sample feasibility chart may be revised as necessary to include any factors that would affect the profit margin of the foodservice outlet. Reprinted with permission of the National Restaurant Association.

divide the annual number of customers into the total costs. Here, the average check would be $4.35.

Using the average check as a guideline, you can now select menu items based on costs of foods, the calculated percentage of each dollar of sales that these costs will represent, and the projected sales mix, or balance of sales of high- and low-profit

```
┌─────────────────────────────────────────────────────────────┐
│  Total estimated operating costs for one year:   1,000,000.00 │
│  20% of total costs as calculated profit margin:    200,000.00 │
│  Total sales needed to cover costs and profit:    2,000,000.00 │
│                                                                │
│  RESTAURANT STATISTICS                                         │
│                                                                │
│  Number of meal services:  2                                   │
│  Anticipated seating turnover                                  │
│       Lunch:    2.5                                            │
│       Dinner:   1.5                                            │
│  Seating Capacity: 250                                         │
│  Number of days open annually: 307                             │
│  Total number of customers served daily: 1,000                │
│  Adjusted average number of customers served                  │
│        daily, based on projected 75% occupancy: 750           │
│  Total estimated number of customers                          │
│        served annually, based on 75% occupancy: 230,250       │
│                                                                │
│  Average lunch check: $6.00                                    │
│     Daily lunch revenue:  469 × 6.00 = $2,814                  │
│                                                                │
│  Average dinner check: $15.50                                  │
│     Daily dinner revenue: 281 × 15.50 = $4,355.50             │
│     Daily lunch and dinner revenue:  $7,169.50                 │
│                                                                │
│  Total estimated annual revenue:          $2,201,036.50        │
│  Total sales needed:                      $2,000,000.00        │
└─────────────────────────────────────────────────────────────┘
```

♦ **FIGURE 2-4.** Worksheet for completing a statement of income and expenses. See Fig. 2-3.

items. Luncheon and dinner menu price ranges are developed by determining an average check for each service. Reviewing the market survey and prices that the direct competition is currently charging, management determines that an average lunch check of $6.00 and an average dinner check of $15.50 will be competitive and profitable. With an average customer count of 469 paying $6.00 each, the daily lunch revenue should be $2,814. The dinner revenue should be $4,355.50 based on a customer count of 281. The combined daily lunch and dinner revenues using those projected customer counts would be $7,169.50. Multiplying this figure by 307 (the number of days open annually) gives a total estimated annual revenue of $2,201,036.50. Total sales needed to realize a profit of $200,000 is $2 million.

The sample feasibility chart shows that this restaurant would be feasible, given the number of customers estimated to be served daily and the predicted operational costs. By calculating the possibility of slow business periods, management has built in a safety cushion that will absorb a certain amount of change. If this cushion is exceeded and the estimated figures change appreciably, however, the profit margin will be narrowed or, more likely, revenues will not exceed costs.

Making a feasibility chart is an important beginning step in planning any business. Unless the costs of operation are estimated, profits planned for, and revenues estimated, management will be operating under severe limitations from a management and business standpoint. In addition, gaining financial support from either banks or investors will be difficult unless an understanding of the financial structure of the restaurant is shown by the success of any foodservice operation. Management and business know-how will make the difference between a profit and a loss, success or failure.

CHAPTER 3

Costing
for
Profit

THE FOUR PARTS OF PROFIT

Before the foodservice manager begins to set menu prices, he or she must determine how much revenue the foodservice organization will have to earn in order to cover all costs and make a profit. Every dollar of sales revenue can be divided into four categories that represent actual costs and projected profits; overhead, labor, food cost, and profit. A breakdown of these costs is outlined in Chapter 2 in relation to determining total estimated operating costs and total sales figure.

In an ideal situation, these costs should represent relative percentages of each dollar of sales totaling 100 percent. The pie chart in Figure 3-1 gives an example of an appropriate percentage breakdown.

Overhead

Overhead includes all the costs connected with operating and maintaining a foodservice location. These costs generally include the following:

Rent or mortgage
Utilities: heat, light, and power
Taxes: federal, state, and local
Licenses
Insurance: property, fire and theft

Overhead costs tend to remain fairly constant from month to month, except for periodic rate increases and fluctuations in utility costs caused by increased or decreased customer business. Overall utility costs can be calculated by forecasting monthly and seasonal increases. Ideally, 20 percent of every dollar in sales should represent overhead.

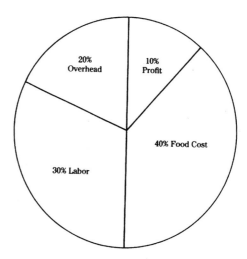

◆ **FIGURE 3-1.** Profit pie-chart showing the suggested percentage breakdown of the four parts of profit.

Labor

The cost of maintaining an operational staff can be broken down into the following five areas:

Payroll
Taxes
Insurance: health, life, and unemployment
Benefits
Personal management and related costs

Thirty percent of every sales dollar should, optimally, represent labor. Changes in labor costs are directly related to minimum-wage increases and are also affected by union negotiations.

Food Cost

Food cost is the total price paid by the establishment for all food items. While the percentage of every sales dollar that represents food costs depends upon the needs of the individual foodservice, food cost should total approximately 35–40 percent of the sales dollar. This percentage represents the overall food cost, not individual relationships between item cost and menu price, which can fluctuate from as low as 18 percent to as high as 60 percent.

Profit

Profit is any money remaining after all expenses have been paid. Operating or gross profit is the total profit before taxes have been subtracted; net profit is the after-tax amount. The profit percentage is completely dependent upon the goals of management and its ability to control costs. A net profit of 10 percent is excellent, but the profit margin in food sales can be slim. Although net profit can be greater than 10 percent on some items, an overall profit of 10 percent is difficult to reach and rarely maintained.

To establish profit goals, each foodservice manager should evaluate current costs in the overhead, labor, and food categories and determine the percentage of total sales that each category now represents. To find this current percentage, the following cost of goods and services exercise can be used (Figure 3-2). Any costs remaining from the previous month are entered in the inventory line (line one) as a carried-over expense. All costs for the month are entered on line two. The total sum of these two figures represents the costs of the goods or services. The total sales figure for that month is then used to determine the percentage that these costs represent of the total revenue for the month.

In the cost of goods and services examples shown in Figure 3-2, a company's cost of food and cost of labor are determined for the month of February. To attain a complete picture of current cost percentages, this company would next establish its overhead percentages for the same month. The cost of goods and services exercise is an excellent one for monthly or periodic review of cost percentages. It should not be used to obtain daily cost fluctuations, however.

```
┌─────────────────────────────────────────────────────────┐
│  COST OF GOODS AND SERVICES EXERCISE                    │
│                                                          │
│  COST OF FOOD                                            │
│  Inventory: January 31                        2,322.76   │
│  Total Food Purchases: February            + 17,500.64   │
│                                               19,823.40   │
│                                                          │
│  Inventory: February 28                     -  3,764.54  │
│  Cost of Food                                 16,058.86   │
│  Total Sales: February                        41,176.56   │
│  Food-Cost Percentage: February                    39%   │
│                                                          │
│  COST OF LABOR                                           │
│  Carried-over Expenses: January                 275.60   │
│  Total Cost of Labor: February             + 12,077.36   │
│  Cost of Labor                                12,352.96   │
│  Total Sales: February                        41,176.56   │
│  Labor-Cost Percentage: February                   30%   │
└─────────────────────────────────────────────────────────┘
```

♦ **FIGURE 3-2.** Cost of goods and services example to determine the percentages that food and labor represent of total sales.

STANDARDS FOR COSTING

Costs of goods in the foodservice industry relate primarily to the food and/or beverage ingredients that go into the menu items sold to the general public. Although we may be able to evaluate profit by comparing the costs of goods against total sales to establish overall gains or losses, this is not an effective daily tool with which to run a successful business. The goals by which production specifications and costs estimates are set must be integrated into daily operations. The selection process of menu items (see Chapter 4) and the development of recipe cards, which will produce a consistent yield and quality of product, provide the basis for determining a standard of profit for each of these menu items. The pie chart in Figure 3-1 shows us how to evaluate all the costs of food against total food sales. An intermediate control step helps to assure that the 40 percent or less allocated to food cost will actually be the percentage of cost over sales at the end of each month. By transferring the specifications on the recipe card onto a standard cost card (Figure 3-3), a standard of costing for each menu item is now established. This is the process that ensures that all food products are purchased and proportioned to yield the exact number of portions designated by the recipe card.

Standards of costing enumerate the actual cost of each ingredient used in a given recipe, to provide a basis for an accurate overall food cost analysis. The standard cost card for each recipe in the menu repertory provides a cost breakdown for each ingredient as well as a total food cost for each recipe. From this information, portion costs can be established along with a final menu price that will enable the foodservice operation to achieve the profit necessary for success. Standards of costing, however, merely reflect the price of maintaining the standards of preparation, production, presentation, and purchasing that have already been established.

Much of the information found on a standard cost card is taken directly from its corresponding standard recipe card. Corresponding index numbers are the same, as

STANDARD COST CARD

ITEM		INDEX #		PORTION YIELD
Ingredient	*Weight*	*Measure*	*CPU*	*Extension*
PC: ABSTRACT: TOTAL PC: FC%: SP:				EXTENSION TOTAL: CF: TOTAL:

◆ **FIGURE 3-3.** A standard cost card is provided for each recipe in the menu repertory. Final portion costs and menu prices can be determined once the food cost for each item is established.

are portions, items, yields, weights, and measures. If an ingredient is used several times within the same recipe, its total weight or measure is listed, rather than each separate weight or measure.

The cost per unit (CPU) is the current wholesale price of the ingredient purchased, and is the basis for determining the extension, or the cost of the amount of the item used in the recipe. To find the extension, either divide the weight or measure into the unit as purchased or multiply the two factors, if the weight or measure is larger than the purchased unit. Before calculating the extensions, however, the weight or measure of the recipe ingredient must be converted to the weight or measure in which the unit was purchased.

Ingredient	Measure	CPU
Strawberries; whole, fresh	1 cup	$0.98 pt

Since the unit of purchase is a pint, convert the measure into pints (1/2) and divide the CPU by that figure: $0.98/2 = $0.49 (extension).

Ingredient	Weight	CPU
Butter	11 oz	$1.89 lb

Here, the extension is determined by converting the pound into 16 ounces and completing the following steps:

$1.89/16 = 0.17 (per oz)

0.17 × 11 (oz) = $0.84 (extension)

The C factor (the cents factor, or CF) represents the cost of ingredients that are valued at a penny or less. The C factor can be found by multiplying the total extension (overall food cost per recipe) by 2 percent. This percentage is an arbitrary one, however, and management may substitute any other percentage that might reflect cost more acurately. Even if the cost card does not have any items that are valued at a penny or less, the C factor is always calculated and added to the cost card. Portion costs are found by dividing the total food costs by the yield. The abstract represents the cost of foods such as condiments, table salt, pepper, sugar, relishes, and so on, which are not part of any standard recipe card. The abstract is generally calculated as 2 percent of the recipe's portion cost.

C factor and abstract costs are added to extension and portion costs to make sure that all costs of food are accounted for and calculated in the selling price of menu items. Although end-of-the-month inventories would show the actual costs of condiments in the total costs of goods, the food cost percentage would be inaccurate because these small costs had not been included in the original pricing.

CONTROLLING MENU COSTS

Once a foodservice organization has established standards and goals in addition to developing systems to ensure consistent costs, quantities, and profits, it is necessary to implement business practices that will assemble the information management needs to keep the operation running at an optimum performance level. Two forms that help the manager carry out day-to-day activities and plan for the future are:

♦ the production sheet
♦ the sales history

The Production Sheet

A production sheet is an itemized list of each food selection to be served at a given meal. It provides an estimate of how many portions of each item will be served and is intended to help the kitchen staff produce the proper amounts of food and to reduce waste and overproduction. To find the total estimated food cost for each item, the estimated number of portions to be served is multiplied by the food cost for each portion.

Estimated portions × food cost per portion = estimated total food cost

65 × $2.24 = $145.60

After the meal service period is completed, the actual number of portions of each item is recorded. For each item, the actual number of portions served is multiplied by the food cost per portion to calculate the actual food cost for the items. So that management can know the total amount of revenue realized for each, the actual number of portions served is multiplied by the selling price of the item, as listed on the menu. The last step is to determine the food cost percentage for each item, which can be found by dividing the actual food cost by the total revenue.

This is the application of the food cost formula covered in Chapter 4. It is also the same process as the cost of goods and services exercise in Figure 3-2.

Actual food cost / total revenue = Food cost percentage

$127.68 / $598.50 = 21%

Once these calculations have been completed, the original estimate of portions to be served is compared to the actual number served. The original estimate can then be evaluated along with an emerging picture of problems with overproduction, underproduction, or planning. To determine the days' overall food cost percentage, the total food cost for the day is divided by the total daily revenue. Foodservice software is available to complete this analysis.

Since it is impossible to estimate the precise number of portions of each item that will be served, waste caused by overprotection must be calculated. First, the actual number of portions served is subtracted from the estimated number of portions served, to determine the number of wasted portions. Next, the number of wasted portions is multiplied by the food cost of each portion, to ascertain the dollar value of wasted portions. To adjust the total revenue received for that item, the dollar value of the waste is subtracted from the orignial revenue received. Last, the actual food cost is divided by the adjusted revenue to determine the percentage of wasted food cost.

The waste percentage calculated from the overprotection is not a true or complete figure in the overall food costs of an operation. The calculation makes no allowance for wasted portions that are reconstituted as ingredients for subsequent menu items held and served at a later meal service, or used as employee meals. In an efficiently managed foodservice operation, waste is planned for and reconstituted in as many ways as possible. The primary reason for calculating waste percentages is to remind management that mistakes in production estimates cost money and to provide the financial impact of monies that would have been lost if these portions were indeed wasted. Although there are often good reasons for an item to sell less than expected, management should bear in mind the costs of errors and the effect on the profit margin.

The sample production sheet shown in Figure 3-4 is categorized by courses. Each item to be produced for a given course would be listed on the production sheet, with a separate production sheet provided for each course. Every foodservice organization has individual needs and problems and should tailor this format to suit their requirements and present production information as clearly as possible. Along with the sales mix and sales history, the production sheet provides information essential to future planning in a foodservice organization.

The Sales History

The sales history is the foodservice operation's daily log. Records are kept of the day's customer counts, total revenue, and overall costs along with weather conditions and any special events that might have affected business. All menu items are listed along with the total number of items sold. Any special problems with personnel or

PRODUCTION SHEET

	TOTAL ESTIMATED CUSTOMER COUNT:					SERVICE:			ACTUAL # SERVED:		
DATE:	TIME:					CHEF:					
ITEM	# OF ITEM TO BE SERVED	FOOD COST	TOTAL FOOD COST	ACTUAL # SERVED	ACTUAL FOOD COST	SELLING PRICE	TOTAL REVENUE	F.C. %	OVERAGE	VALUE	WASTE F.C. %

◆ **FIGURE 3-4.** The production sheet provides an estimate of the total number of each food item to be served at a given meal.

SALES HISTORY

DATE:		DAY:		TIME:		BY:	
WEATHER CONDITIONS:							
SPECIAL EVENTS OR PROBLEMS:							
# OF COVERS:				# OF CHECKS:			
Item	*# Sold*	*Sales*		*Item*	*# Sold*	*Sales*	
BEVERAGES *Wine* *Beer* *Liquors* *Other:*							

◆ **FIGURE 3-5.** Sample sales history format. The sales history is the restaurant's daily log of customer counts, total revenues, and costs, along with any special factors that might have affected sales.

customers are also noted. In an operation with more than one outlet, a separate sales history is kept for each outlet.

Over time, the sales history provides valuable information concerning past successes and failures of the foodservice organization. It offers the data needed to make up production sheets and staffing schedules in addition to supplying the information needed to complete the sales mix. A suggested sales history format is shown in Figure 3-5. As with any sample form, each outlet should review the format and change the design to meet specific requirements.

CHAPTER 4

♦

Pricing
for
Profit

Menu pricing—establishing sales prices for menu items—can be approached from a number of different perspectives. Each foodservice manager has individual requirements and circumstances that determine the basic formulas and methods used to establish prices. Although several common menu pricing theories and formulas are presented in the following section, it is important to note that each foodservice operation is unique and needs to be evaluated individually. In adopting a formula that meets the goals and requirements of both management and customer, two basic factors should be considered:

1. How much revenue is needed to cover costs and provide a given profit?
2. What is the perceived value that customers have for these menu items? What will the customer find is an acceptable price?

There are four generally accepted pricing formats in which food selections are presented:

♦ table d'hote (fixed price)
♦ semi table d'hote
♦ a la carte
♦ semi a la carte

These different pricing formats and their applications to different types of restaurants are discussed in the following sections. Each form addresses a particular need on the part of the customer as well as the restaurant. These four menu formats can be used alone or combined to produce a varied presentation of food items and pricing systems.

The original American menu format was table d'hote (literally, the regular meal of the host). In this format, one entire meal is offered at a fixed price (prix fixe is the French term). While some menus may present different items to choose from per course, one total price is applied to the entire meal. This pricing method was adopted in America by small cities and towns where the local hotel or inn was the only restaurant. As guests had no alternative public dining room, the hotel's practice of offering one set meal for lunch and dinner was acceptable. Figure 4-1 shows a table d'hote menu from the Hotel Vernet, Paris. The luncheon menu offers a choice of entrees because local business people as well as hotel guests patronize the hotel for the midday meal. The dinner menu limits the entree to one selection, as dinner customers are usually limited to hotel guests and the use of an a-la-carte menu would be inappropriate, because the customer has already paid for meals in the daily or package per-person charge.

Current use of table d'hote is generally limited to resort hotels, cruise ships, private banquets, and other situations in which the customer is more or less "captive." The example of a cruise ship menu in Figure 4-2 shows a variation on the table d'hote format. The menu offers two plans—a seven-course menu and a "healthy" four-course menu. A main course option for vegetarians can be substituted for either menu, with the guest selecting cream of mushroom rather than chicken soup. Wine is also included for this meal. Commercial restaurants frequently use table d'hote as the format for special-occasion and holiday menus such as the Easter brunch menu in Figure 4-3.

◆ FIGURE 4-1. Table d'hôte menu from the Hotel Vernet, Paris.

◆ FIGURE 4-2. Table d'hôte menu featuring both a standard menu and healthy dining menu from Radisson Cruise Lines.

◆ **FIGURE 4-3.** Table d'hôte menu for commercial restaurant use as an Easter menu from the Hotel duPont, Wilmington, Delaware. Reprinted with permission of the Hotel duPont, Wilmington, Delaware.

Perhaps the most common use of table d'hote in commercial foodservice is in banquet or catering menus. The extent to which banquet or catering is developed as an outlet for a foodservice operation will determine the selection of menu items offered. Hotels, restaurants, and catering houses that provide banquets as a significant part of their business will offer menus to suit any formal or informal occasion. Banquet menus can be offered in any of the four forms discussed, but the most popular are semi a la carte and table d'hote. With table d'hote, the menu is completely planned, proportioned, and prepriced, based on an established per-person consumption. The customer is offered a number of table d'hote menus within a given price range from which to choose. Items and prices may be tailored to accommodate customer needs and desires. The banquet menu in Figure 4-4 is a three-course table d'hote dinner menu. A optional beverage package is noted at the bottom of the menu.

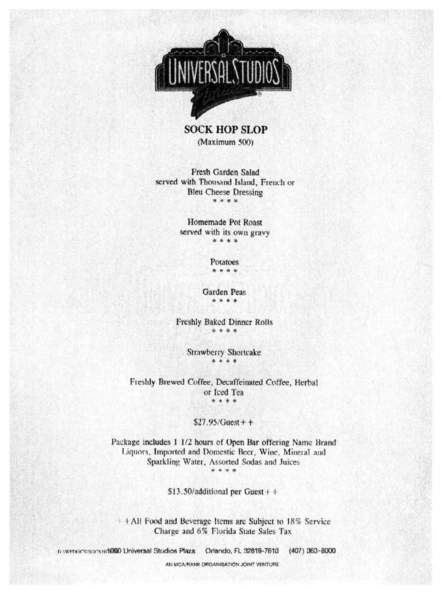

SOCK HOP SLOP
(Maximum 500)

Fresh Garden Salad
served with Thousand Island, French or
Bleu Cheese Dressing
* * * *

Homemade Pot Roast
served with its own gravy
* * * *

Potatoes
* * * *

Garden Peas
* * * *

Freshly Baked Dinner Rolls
* * * *

Strawberry Shortcake
* * * *

Freshly Brewed Coffee, Decaffeinated Coffee, Herbal
or Iced Tea
* * * *

$27.95/Guest + +

Package includes 1 1/2 hours of Open Bar offering Name Brand
Liquors, Imported and Domestic Beer, Wine, Mineral and
Sparkling Water, Assorted Sodas and Juices
* * * *

$13.50/additional per Guest + +

+ + All Food and Beverage Items are Subject to 18% Service
Charge and 6% Florida State Sales Tax

1000 Universal Studios Plaza   Orlando, FL 32819-7610   (407) 363-8000
AN MCA/RANK ORGANISATION JOINT VENTURE

♦ **FIGURE 4-4.** A variation of the table d'hôte menu used as a commercial banquet menu from Universal Studios, Florida. Reprinted with permission of Universal Studios, Florida.

As the availability of restaurants grew in the United States, so did the competition. Customer demand triggered the development of a menu form called semi table d'hote. In this system, one or two courses are priced separately, and the rest of the courses are treated as one meal and given one price, as in the table d'hote format. The appetizer and dessert courses are frequently priced separately, since these are the courses that customers generally prefer to eliminate. A semi table d'hote menu is shown in Figure 4-5 combined with a la carte sandwich prices.

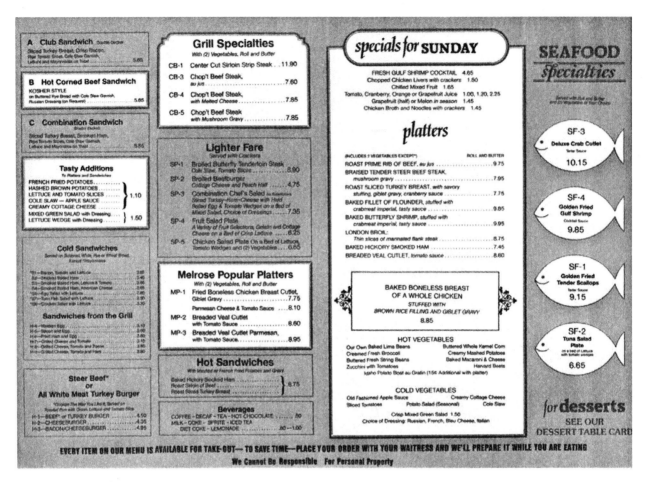

◆ FIGURE 4-5. A combination of a la carte, semi table d'hôte and table d'hôte pricing are featured in this menu from the Melrose Diner in Philadelphia, Pennsylvania. Reprinted with permission of Melrose Diner, Philadelphia, Pennsylvania.

The traditional a la carte menu from Llangoled Hall in Wales, Great Britain, shown in Figure 4-6, lists each of the menu items in order of the standard sequences of courses, as discussed on page 5, Figure 1-3. The basic principle of the a la carte menu form is that every item is priced separately. It is up to the customer to make selections and the customer is charged only for the items ordered.

The popular semi a la carte menu format was developed in the United States. In this form, a limited number of courses are included in the price of the entree while the remaining courses are priced separately. Depending on the restaurant, the main course price could include the entree, vegetable, starch, and salad, with the appetizer, soup, dessert, and beverage priced individually.

The European customer has long been accustomed to paying for each course separately, but American customers have developed buying habits from the all-inclusive perspective. The concept of paying one price for the main course and being "given" the surrounding items that would normally be the accompaniments convinces

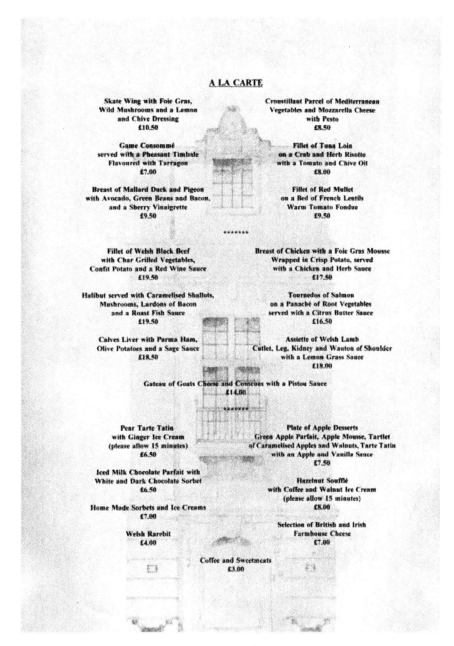

A LA CARTE

Skate Wing with Foie Gras,
Wild Mushrooms and a Lemon
and Chive Dressing
£10.50

Croustillant Parcel of Mediterranean
Vegetables and Mozzarella Cheese
with Pesto
£8.50

Game Consommé
served with a Pheasant Timbale
Flavoured with Tarragon
£7.00

Fillet of Tuna Loin
on a Crab and Herb Risotto
with a Tomato and Chive Oil
£8.00

Breast of Mallard Duck and Pigeon
with Avocado, Green Beans and Bacon,
and a Sherry Vinaigrette
£9.50

Fillet of Red Mullet
on a Bed of French Lentils
Warm Tomato Fondue
£9.50

****

Fillet of Welsh Black Beef
with Char Grilled Vegetables,
Confit Potato and a Red Wine Sauce
£19.50

Breast of Chicken with a Foie Gras Mousse
Wrapped in Crisp Potato, served
with a Chicken and Herb Sauce
£17.50

Halibut served with Caramelised Shallots,
Mushrooms, Lardons of Bacon
and a Roast Fish Sauce
£19.50

Tournedos of Salmon
on a Panaché of Root Vegetables
served with a Citrus Butter Sauce
£16.50

Calves Liver with Parma Ham,
Olive Potatoes and a Sage Sauce
£18.50

Assiette of Welsh Lamb
Cutlet, Leg, Kidney and Wanton of Shoulder
with a Lemon Grass Sauce
£18.00

Gateau of Goats Cheese and Couscous with a Pistou Sauce
£14.00

****

Pear Tarte Tatin
with Ginger Ice Cream
(please allow 15 minutes)
£6.50

Plate of Apple Desserts
Green Apple Parfait, Apple Mousse, Tartlet
of Caramelised Apples and Walnuts, Tarte Tatin
with an Apple and Vanilla Sauce
£7.50

Iced Milk Chocolate Parfait with
White and Dark Chocolate Sorbet
£6.50

Hazelnut Soufflé
with Coffee and Walnut Ice Cream
(please allow 15 minutes)
£8.00

Home Made Sorbets and Ice Creams
£7.00

Selection of British and Irish
Farmhouse Cheese
£7.00

Welsh Rarebit
£4.00

Coffee and Sweetmeats
£3.00

♦ **FIGURE 4-6.** A traditional a la carte format is used in this menu from Llangoled Hall, Wales, Great Britain. Each menu item is priced separately. Reprinted with permission of Ashley Hotels.

the customer that they are getting something extra. Although we all know that the price of the entree is calculated to include surrounding items, the psychology of one price for all is very successful with the American customer market.

Many restaurants have successfully combined one or all of these menu formats to meet customer demand. In the breakfast menu in Figure 4-7, three of the pricing

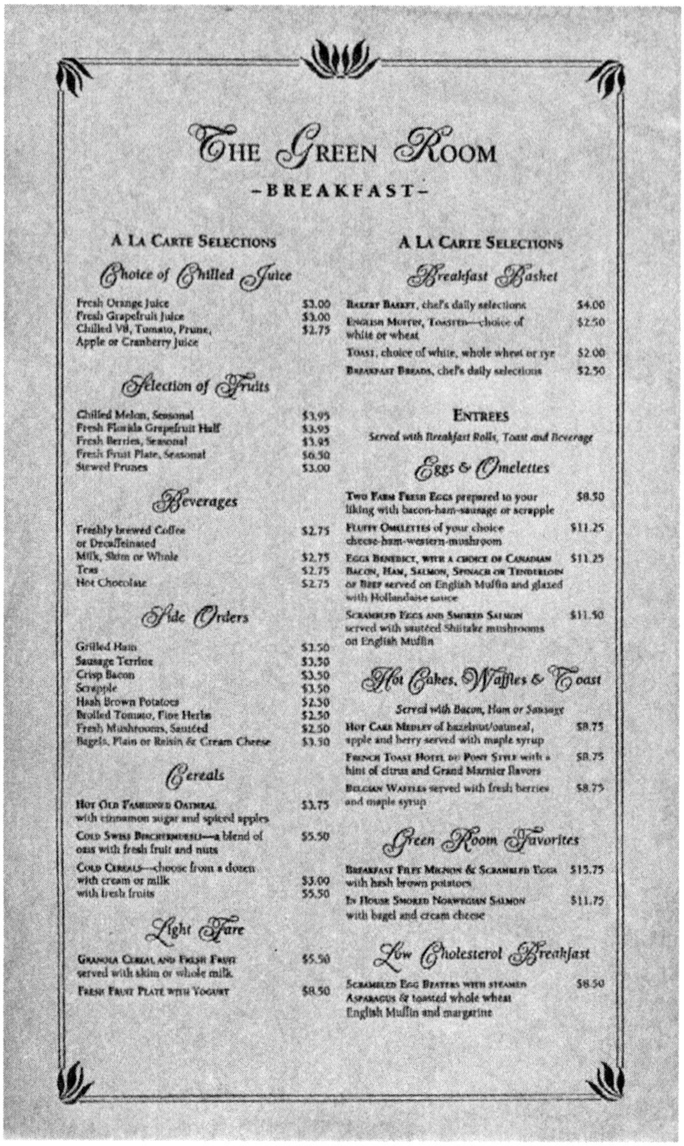

THE GREEN ROOM

-BREAKFAST-

A LA CARTE SELECTIONS

Choice of Chilled Juice

Fresh Orange Juice $3.00
Fresh Grapefruit Juice $3.00
Chilled V8, Tomato, Prune, Apple or Cranberry Juice $2.75

Selection of Fruits

Chilled Melon, Seasonal $3.95
Fresh Florida Grapefruit Half $3.95
Fresh Berries, Seasonal $3.95
Fresh Fruit Plate, Seasonal $6.50
Stewed Prunes $3.00

Beverages

Freshly brewed Coffee or Decaffeinated $2.75
Milk, Skim or Whole $2.75
Teas $2.75
Hot Chocolate $2.75

Side Orders

Grilled Ham $3.50
Sausage Terrine $3.50
Crisp Bacon $3.50
Scrapple $3.50
Hash Brown Potatoes $2.50
Broiled Tomato, Fine Herbs $2.50
Fresh Mushrooms, Sautéed $2.50
Bagels, Plain or Raisin & Cream Cheese $3.50

Cereals

Hot Old Fashioned Oatmeal with cinnamon sugar and spiced apples $3.75
Cold Swiss Birchermuesli—a blend of oats with fresh fruit and nuts $5.50
Cold Cereals—choose from a dozen
with cream or milk $3.00
with fresh fruits $5.50

Light Fare

Granola Cereal and Fresh Fruit served with skim or whole milk $5.50
Fresh Fruit Plate with Yogurt $8.50

A LA CARTE SELECTIONS

Breakfast Basket

Bakery Basket, chef's daily selections $4.00
English Muffin, Toasted—choice of white or wheat $2.50
Toast, choice of white, whole wheat or rye $2.00
Breakfast Breads, chef's daily selections $2.50

ENTREES

Served with Breakfast Rolls, Toast and Beverage

Eggs & Omelettes

Two Farm Fresh Eggs prepared to your liking with bacon-ham-sausage or scrapple $8.50
Fluffy Omelettes of your choice cheese-ham-western-mushroom $11.25
Eggs Benedict, with a choice of Canadian Bacon, Ham, Salmon, Spinach or Tenderloin of Beef served on English Muffin and glazed with Hollandaise sauce $11.25
Scrambled Eggs and Smoked Salmon served with sautéed Shiitake mushrooms on English Muffin $11.50

Hot Cakes, Waffles & Toast

Served with Bacon, Ham or Sausage

Hot Cake Medley of hazelnut/oatmeal, apple and berry served with maple syrup $8.75
French Toast Hotel du Pont Style with a hint of citrus and Grand Marnier flavors $8.75
Belgian Waffles served with fresh berries and maple syrup $8.75

Green Room Favorites

Breakfast Filet Mignon & Scrambled Eggs with hash brown potatoes $15.75
In House Smoked Norwegian Salmon with bagel and cream cheese $11.75

Low Cholesterol Breakfast

Scrambled Egg Beaters with steamed Asparagus & toasted whole wheat English Muffin and margarine $8.50

♦ **FIGURE 4-7.** This breakfast menu from the Hotel duPont in Wilmington, Delaware combines a la carte and table d'hôte pricing concepts. Reprinted with permission of Hotel duPont, Wilmington, Delaware.

formats are combined, from a complete table d'hote breakfast to semi a la carte to a la carte selections.

The ability of the menu planner to calculate customer needs and desires in pricing will determine the success of the menu's selling power. Accurate records of customer buying trends and reactions should be kept. In developing a menu pricing concept for a new restaurant, the competitions menu formats should be thoroughly analyzed as well.

♦ **FIGURE 4-8.** An a la carte menu from Pusser's at the Beach, Ft. Lauderdale, Florida.

The full-service dinner menu shown in Figure 4-8 is an excellent example of almost 100 percent a la carte pricing, as it has become standard in most American restaurants. The fast-food boom of the 1970s and 1980s helped to make this pricing concept more acceptable to the American restaurant customer. Fast food restaurants provide both a la carte and semi a la carte pricing. By selling items both individually and in "bundles" of discount priced items, operators promote special items and increase revenues by encouraging customers to buy more than they would originally have in a package price.

MENU PRICING METHODS

Because foodservice organizations have different ways of operating, several different methods can be used to assign menu prices. Five methods and their formulas are discussed in this section:

♦ Food cost percentage method
♦ Factor pricing method
♦ Prime cost method
♦ Actual cost method
♦ Contribution to profit method

Food Cost Percentage Method

The food cost percentage method of menu pricing is based on the theory that the three categories of cost (overhead, labor, and product) should occupy fixed percentages of each revenue dollar, so that a fixed percentage of profit can be pre-achieved. As mentioned earlier, the percentage breakdown of each revenue dollar would be as follows:

- Overhead   20%
- Food Coast   40%
- Labor   30%
- Profit   10%

This percentage breakdown is, however, only a suggested one. Each foodservice operation should adjust the percentages to suit their specific needs. Many foodservice outlets, both commercial and institutional, choose to adopt significantly higher or lower food cost percentages. Whichever proportion or ratio is chosen, a sufficient percentage should be allotted to maintain food quality and quantity while allowing enough revenue for labor, overhead, and profit. Three factors are involved in any food cost percentage problem: the cost of food, the selling price, and the desired food cost percentage. Two of the factors must be known in order to solve for the other factor. For example, to determine the selling price of a menu item, divide the food cost of the item by the desired food cost percentage. Table 4-1 illustrates how to solve for each of the three factors if the other two are known. By assigning a value to each cost, a budget is determined for each cost and guidelines for overall cost controls are established.

Factor Pricing Method

The factor pricing method is essentially a simplified version of the food cost percentage method. First, the percentage that food cost will represent of each revenue dollar is determined. The percentage is then converted into a factor that will always represent

---

♦ TABLE 4-1 ♦

Determining a Selling Price

Unknown	Known Factor
Selling Price (SP) =	FC/FC%
Food cost percentage (FC%) =	FC/SP
Food Cost (FC) =	SP × FC%

♦ 4-2 ♦

Factor Pricing Method

1. The selling price of the menu item is $5.00. To find a factor that will represent 25 percent of $5.00, divide 100 by 25.

$$100\%/25 = 4$$

$$\$5.00/4 = \$1.25$$

2. The selling price of a menu item is $7.00. To find a factor that will represent 35 percent of $7.00, divide 100 by 35.

$$100\%/.35 = 2.8$$

$$\$7.00/2.8 = 2.50$$

the food cost percentage, and the factor is divided into the selling price to equal the actual food cost to the foodservice operation. This method is illustrated in Table 4-2.

Prime Cost Method

The prime cost method of menu pricing shown in Table 4-3 is used when a foodservice organization buys a substantial number of preprepared food items. Preprepared foods, which are purchased in a finished or nearly finished state, can be served immediately or completed with a minimum of labor. By using the prime cost method, the foodservice manager can identify what percentage of the cost of the preprepared food items represent production labor. In the present market, production labor is a significant issue in maintaining a balance of quality food items against rising labor

♦ TABLE 4-3 ♦

Prime Cost Method

Menu Item	Food Cost	Labor Cost
Prepared:		
Breast of Capon	1.15	0.70
Green Beans	0.20	0.10
Raw Food:		
Sauce Veronique	0.22	0.20
Rice Pilaf	0.28	0.08
	1.85	1.08

costs and a shortage of mid-level kitchen help and preparation staff. An excellent example of these considerations is seen in Chapter 6, where the difference between the actual yield of breaking down a side of beef and buying precut, proportioned cuts of the same quality of beef are compared.

When calculating the prime cost method, first classify every item on the menu as either raw or prepared. In-house production time is determined for all raw items and a preportioned labor cost assigned by giving an overall labor cost to the recipe and dividing it by the number of portions. Any production costs incurred in the completion of the preprepared items must also be determined and assigned. To calculate production costs for preprepared foods, itemize each preparation task, determine how much time is required to complete the task, and then assess the cost of the labor time involved. In Table 4-3, for example, the labor cost of $0.55 for the breast of capon reflects the value of the time required to place the capon on a sheet pan, put it in and take it out of the oven, and arrange it on a plate. Last, a selling price that includes some food and labor costs is calculated, using the food cost percentage method discussed earlier. This calculation will alter the percentage ratios of food cost and labor in relation to the amount of overall labor that has been included in the production calculations.

Actual Cost Method

The actual cost method, as shown in Figure 4-9, is used when management needs to know how much to allocate for food cost, so that menu items can be selected and prices calculated. In this approach, profit is determined before the total desired revenue is established. This situation would occur, for instance, if investors in a foodservice operation had predetermined the dollar amount of profit that they wished to realize.

The profit is given a dollar value that represents a fixed percentage of that total revenue. Overhead and labor are also given dollar values, based on known costs and fixed or estimated percentages. These three costs are totaled and subtracted from the balance; the remainder represents food cost, from which the cost of menu items will be determined. Instead of assigning menu prices to a final list of items, however, menu items are selected or dropped based on competitive offerings of other restaurants (as seen in the market survey), management's menu selection, and the amount of money that can be spent on food for each item. To find the amount of money available for each item, divide the annual desired revenue by the forecasted annual customer count. This amount is the food cost available for the average check. At this point, management determines how many and what kinds of menu items will be included in the average check. For example, if the average check was $19.00, as shown below, and included an appetizer, entree, and dessert, the following breakdown of item costs might be applied. Appetizer and dessert are about 15 percent and 14 percent of the check; the entree, roughly 70 percent.

Appetizer	$3.00 × 0.40 FC% =	$1.20
Entree	13.00 × 0.40 FC% =	5.20
Dessert	2.50 × 0.40 FC% =	0.90
	$19.00	$7.30

```
┌─────────────────────────────────────────────────────────────────┐
│ ACTUAL COST METHOD                                                │
│                                                                   │
│ 1. Predetermined Annual Profit (10%):              $200,000.00    │
│ 2. Desired Annual Revenue:                        2,000,000.00    │
│        10% Profit:                                  200,000.00    │
│        20% Overhead:                                400,000.00    │
│        30% Labor:                                   600,000.00    │
│        Total Costs:                               1,200,000.00    │
│ 3. Amount available for food cost, after                          │
│    subtracting total costs from revenue:            800,000.00    │
│ 4. Average check, after dividing total                            │
│    revenue by estimated annual customer                           │
│    count:                                               16.00     │
│                                                                   │
│       Total revenue    2,000,000.00                               │
│       ───────────── =  ──────────── = $16.00                      │
│       Customer count      125,000                                 │
│                                                                   │
│ 5. Food cost available for each average                           │
│    check, after dividing annual funds                             │
│    available for food by estimated annual                         │
│    customer counts:                                      6.40     │
│                                                                   │
│       Annual food cost   800,000.00                               │
│       ──────────────── = ────────── = $6.40                       │
│       Customer count       125,000                                │
│                                                                   │
│ 6. Food-cost percentage, after dividing                           │
│    annual funds available for food by total                       │
│    revenue:                                              40%       │
│                                                                   │
│       Annual food cost  800,000.00                                │
│       ──────────────── = ──────────── = 40%                       │
│       Total revenue     2,000,000.00                              │
│                                                                   │
└─────────────────────────────────────────────────────────────────┘
```

♦ **FIGURE 4-9.** The actual cost method establishes how much money is available in a restaurant's budget for the cost of food. This method is used to pre-establish profit.

By using the desired food cost percentage formulae discussed earlier (FC = SP × FC%), the desired food cost of 40 percent is multiplied by the selling price to obtain the food costs of the three items. The exact percentage of the average check that each menu item will represent will depend on the final pricing structure of the entire menu.

While the ability to establish a standard food cost for the menu is ideal and looks good on paper, a variable factor that enters into the success of this pricing theory is the customer's perceived value of a menu item. Although an appetizer may need to be priced at $7.50 in order to maintain its proper position in food cost control, customers may not be willing to pay more than $5.50 based on the value they assign to the them. On the other hand, an entree priced at $9.50 may actually have a higher perceived customer value and be able to receive a menu price of $11.25. The sales mix shows that the average number of $5.00 appetizers sold is 25 per meal service, while the main course sales are calculated at 45 per meal service. Using this pricing theory, the main course gains $1.75 (9.50 + 2.75 = 11.25) or $78.75 in additional profit per service. The appetizer item losses $2.50 per item (7.50 − 2.50 = 5.00), and $62.50 per meal service. When the numbers on both items are compared, the restaurant has net gain of $16.25 ($78.25 − $62.50 = $16.25). This method of evaluating the popularity index of menu items is one of the major methods of adjusting the operations menu prices to meet customer needs. In order to make this an effective costing method, sales mixes must be accurate and monitored on a weekly basis.

Contribution to Profit Method

The contribution to profit method of pricing, which is very similar to the actual cost method, supposes that a set range of menu prices exists based on current competition and customer demand. Table 4-6 identifies each menu item according to its actual food cost and food cost percentage of the established selling price. The selling price is then compared to the actual menu price and the item's contribution to the gross profit of the restaurant is calculated.

Item A has a food cost of $4.55 and a selling price of $13.67 based on a 33 percent desired food cost. The adjusted menu price if $13.95 and the overall amount of gross profit that this item will contribute to the daily receipts if $9.40. Item B, on the other hand, has a total cost of $2.62 and a selling price of $7.86. The customer's perceived value for this item, however, will result in an accepted menu price of $10.95. This marketing strategy creates a contribution to gross profit of $8.33 per item. Item C balances in the other direction by costing out at $6.12 with a selling price of $18.36. Management knows that the customer will resist this high a price and drops back to $15.95 with an adjusted 38 percent food cost. The contribution to gross profit is still a healthy $9.83 per item. The success of this combined accounting and marketing pricing method requires the same constant monitoring as the actual cost method. The daily sales mix shown in Figure 4-4 is the most accurate method of recording information needed to evaluate these types of pricing methods. Table 4-4 establishes the popularity index of menu entrees over an 18-day period as it might appear on the sales mix report. The average number of sales per day determines the position. This information is then transfered onto the table

♦ TABLE 4-4 ♦

Sales Mix Popularity Index

Entree	18-Day Sales	Average Number per Day	Popularity Index (%)
Veal Parmigiana	698	39	13
Filet Oscar	901	50	16
Prime Rib	967	54	18
Shrimp Scampi	463	26	9
Salmon Steak	261	15	5
Scallops	492	27	9
Swordfish	403	25	8
Lamb Chops	257	14	5
Chicken Forestiere	137	7	2
Capon Breast	**834**	**46**	15
	5413	303	

♦ 4-5 ♦

Contribution to Profit

Item	Food Cost (FC)	(FC%)	Selling Price	Menu Price (MP)	Contribution Gross Profit (CGP)
A	$4.55	33	$13.67	$13.95	$9.40
B	$2.62	33	$7.86	$10.95	$8.33
C	$6.12	33	$18.36	$15.95	$9.83

shown in Table 4-5 to determine the average contribution to profit for this item. If the sales history records that the contribution to profit is consistent for a 9–12 month period, management can be reasonably sure of the volume of profit being maintained for this item. The success of the process is, of course, dependent on all of the menu controls and proper record keeping being practiced. Foodservice management software programs tabulate the information for management to analyze.

CATERING MENU PRICING

Calculating Selling Price

The theories of the menu costing and pricing formats discussed in this section are easily applied to catering menus and packages. When catering is an extension of a restaurant menu, it is important to take into account equipment that may need to be rented to supplement on-site inventory. If items from the restaurant menu are being sold for service in private function rooms, customers will expect to be charged the same prices. Management must take into account the total costs for catered functions, adding a service charge if appropriate.

Ideally, food cost should be between 17 percent and 27 percent for volume service. Volume purchase prices and production techniques can be taken advantage of as well as reduced labor costs. Giving added value to the customer in the form of additional courses and food items often creates a perceived value for catering menus priced higher than restaurant menus.

Successful catering menu pricing requires that management develop specifications for each menu item. As seen in Figure 4-10, specifications detail portion size, source of product (fresh or canned), sizing (as in baby shrimp), and recipe designation (New England/Manhattan). A cost is then determined for a single portion of each menu item. By costing menu items separately, such as the appetizer section, management can decrease or increase the selling price of a catering menu based on customer preferences. A catering menu, for example, may offer Manhattan Clam Chowder and a Caesar Salad with a Stuffed Breast of Chicken Borsin. The customer asks for a fresh fruit appetizer, preferably melon and a salad of mixed greens. From the

Catering Menu Item Specifications

Appetizers

Menu Item	Portion Size	Cost Per
Fruit Cup	6 oz mixed fresh and canned fruit	.93
Melon	1/2 fresh seasonal melon	1.02
Marinated Shrimp	6 oz baby shrimp papaya slice	1.10
Consomme en Tasse	6 oz consomme 2 cheese straws	.51
Clam Chowder	6 oz New England	.37

Salad

Ceasar Salad	6 oz greens with dressing	.59
Marsailles Salad	6 oz mixed greens with dressing	.72
Marinated Vegetable	7 oz mixed vegetable	.90

Entrees

Stuffed Breast of Chicken/Boursin	8 oz breast stuffed	3.75
Poached Salmon	6 oz fresh	3.20
Prime Rib	12 oz prime rib	5.28
Roast Sirloin of Beef sliced	8 oz sirloin wild mushroom sauce	3.60

Dessert

Chocolate Mousse	3 oz chocolate mousse whipped cream	.55
Key Lime Pie	1/6 slice of pie	.80
Bombe Marie Louise	frozen bombe	.83

◆ **FIGURE 4-10.** The catering menu item specification form details portion size, product source, and cost per item.

Catering Menu Specification Form shown in Figure 4-10, management now selects appropriate menu items for a total food cost of 29% of $25.00 as seen in Table 4-6.

If the detailed menu and price are acceptable to the customer, management has a choice of maintaining the 29% food cost in order to balance off other functions that may run above the desired food cost percentage, or of upgrading items such as

♦ **TABLE 4-6** ♦

Establishing Final Food Cost % Based
on Menu Item Costs

Menu Item	Cost	Selling Price
		$25.00
Fresh Fruit	$ 0.93	1.02
Salad	0.72	
Entree/Salmon	3.02	
Vegetable	0.60	1.30
Starch/Rice	0.30	
Roll and Butter	0.30	
Dessert	0.80	1.50
Beverage	0.40	
Total food cost:	7.25	8.74
Selling Price:	$25.00	
Food Cost %	29% FC	35% FC

dessert and appetizer. By increasing the total food cost, the food cost percentage will be 35% and the remaining $1.50 for food cost can be accounted for.

Catering Menu Pricing Formats

Catering menus generally use the table d'hote or fixed price, method to establish a selling price, as in the menu in Figure 4-11. Variations on this format allow for the customer to upgrade one or more sections of the menu, such as dessert, and increase the menu price, as shown in Figure 4-12.

A la carte catering menu pricing has been gaining in popularity, particularly in metropolitan areas. Figure 4-13 offers a complete a la carte menu to the catering customer. Prices are printed on a separate page to reduce the cost of menu printing. The potential problem with this format is that customers will put together unsuitable combinations of food items, both from a presentation and an operational standpoint.

Maintaining Fixed Percentages

The importance of maintaining established food cost percentages can be seen in these methods of menu pricing. The cost of goods and services exercise discussed at the beginning of this section is an excellent one for periodic evaluations of costs, but does not reflect the daily cost fluctuations of individual items nor reveal any additional problems that might drastically affect the month-end percentages. Food cost percentages can be affected by waste, improper portioning, theft, and other factors. If a problem began occurring on the fifth of February and the food cost percentage was not reviewed until the twenty-eighth of the month, a considerable amount of money could be lost. For example, a line cook may have used a combination

---

LUNCHEON

Caesar Style Salad

(Crisp Romaine Lettuce , Fresh Grated Parmesan Cheese
with Roasted Croutons and Caesar Style Dressing)

Chicken Brochette

(Tender Marinated Chicken with Onions, Cherry Tomatoes
and Grilled Red & Green Peppers, served over Rice)

Fresh Breads and Butter

Tropical Torte
with Raspberry Sauce

Coffee, Tea and Iced Tea

Per Person: $ 21.75

Prices do not include 19% Service Charge or Applicable State Tax

---

♦ **FIGURE 4-11.** A basic catering luncheon menu with table d'hôte pricing.

of oversized plates and serving platters when plating up items during the lunch service. Because the items were dwarfed by the plates, the cook added extra food and garnishes not originally specified. Unfortunately, these extras add up to approximately 40 cents in overportioning costs. If the number of lunch plates served was 175, this loss would total $70 per day. If the practice continued unchecked for 23 days, the value of the loss would total $1,610. The total loss would be significant and the situation would not be corrected in time to do anything about profits for the month of February.

A daily food cost analysis can be calculated by totaling the value of that day's food requisitions and dividing the value by the total food sales for the same day. If done by hand, this process is time consuming, as each food requisition must be reviewed, but having the up-to-date food cost percentages justifies the effort. Foodser-

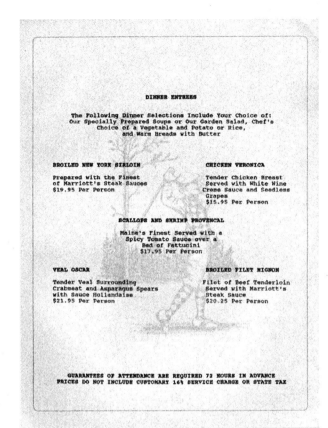 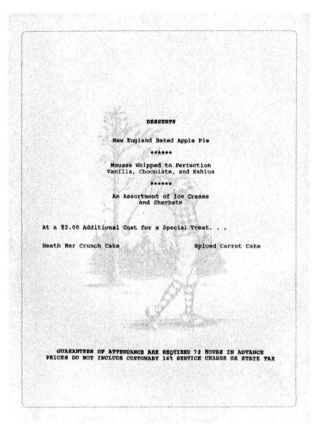

♦ **FIGURE 4-12.** This catering menu from the Portland, Maine Marriott offers the customer the ability to upgrade specific items on the table d'hôte menu. Reprinted with permission of Portland Marriott at Sable Oaks.

vice software programs eliminate a significant amount of time and labor involved in such calculations.

MARKETING TECHNIQUES FOR SUCCESSFUL PRICING

As noted previously, foodservice managers must take the customer's perceived value of an item into consideration when establishing prices. No matter how well menu prices reflect cost and profit percentages, the customer simply may not like the price and decide not to buy the product. To minimize this, marketing techniques should be applied to menu prices. The techniques discussed in steps 1–5 below are guidelines for developing menu prices:

Step 1. A bottom-line selling price for each item on the menu should be established, using one of the menu pricing methods: actual cost, food cost percentage, prime cost or factor pricing.

CHILLED APPETIZERS

Smoked Norwegian Salmon, Marinated Fennel, Caper Sauce

Prosciutto di Parma, Roasted Peppers, Tomato Oil

Seared Eggplant and Goat Cheese Terrine, Red Pepper Coulis

Vine Ripened Roma Tomatoes, Mozzarella Cheese, Basil Vinaigrette

HOT APPETIZERS

Cappellini Pasta, Asparagus, Toasted Pine Nuts, Basil Cream Sauce

Penne Pasta, Prosciutto, Peppers, French Beans, Tomato Vodka Sauce

Baked New England Cod, Bay Shrimp, Nantua Sauce

Artichoke Bottoms, Herb Boursin, Tomato Fondue

Seared Gulf Shrimp, Grilled Polenta, Cilantro Sauce

SOUPS

The Ritz-Carlton Boston Clam Chowder

Chicken Consomme, Julienne of Vegetables, Tarragon Spaetzle

Vegetable, White Bean, Minestrone, Grated Parmesan

Anaheim Pepper, Tortilla Soup

Chilled Gazpacho with Tomatillos

Lobster Bisque

SALADS

Traditional Caesar

American Field Greens, Champagne Vinaigrette

Heart of Romaine, Marinated Mushrooms, Red Pepper, Balsamic Vinaigrette

Marinated Seared Beef Tenderloin, Baby Greens, Black Pepper Vinaigrette

Bibb Lettuce, Enoki Mushrooms, Papaya, Pine Nuts, Papaya Seed Vinaigrette

Endive, Radicchio, Mizuna, Bibb Lettuce
Asparagus, Tomato, Lemon Mustard Vinaigrette

California Baby Greens, Peppered Goat Cheese
Orange Segments, Citrus Vinaigrette

A 19% Service Charge and Applicable State Sales Tax on Food and Beverage will be added. Prices are subject to change without notice.

CHILLED ENTREES

Spinach Salad, Smoked Chicken Breast, Walnuts, Granny Smith Apples, Cider Vinaigrette

Muffalata with Smoked Turkey, Genoa Salami, Provolone, Lettuce, Tomato, Bow Tie Pasta Salad

Grilled Chicken Breast, Louisiana Bay Shrimp Salad, Roma Tomatoes
Bibb Lettuce and Marinated Hearts of Palm

Romaine Cobb Salad, Blue Cheese or Mustard Dressing

Almond Chicken Salad served on Croissant, Seasonal Fruit, Pasta Salad

Grilled Ahi Tuna on Caesar Salad, Cherry Tomatoes and Nicoise Olives

Poached Atlantic Salmon, Marinated Cucumber Salad, Asparagus, Citrus Vinaigrette

MEAT AND POULTRY

Grilled Chicken Breast on Pesto Linguini, Herb Chardonnay Sauce

Chicken Breast Stuffed with Wild Mushrooms and Leeks, Red Pepper Cream

Sauteed Chicken Breast, Sauvignon Blanc Cream with Bay Shrimp, Pistachios and Capers

Sauteed Veal Loin, Prosciutto and Spinach on Oregano Linguini, Asiago Cream

Cannelloni of Ricotta Cheese, Chicken and Spinach, Alfredo and Marinara Sauce

Veal Loin Stuffed with Asparagus and Sun-dried Tomatoes, Pinot Noir Sauce

Sauteed Veal Medallions, Plum Tomatoes, Sage and Prosciutto Relish, Tarragon Sauce

Grilled Filet of Beef, Vegetable Linguini and Wild Mushrooms, Madeira Sauce

Grilled New York Steak and Blackened Gulf Shrimp, Cilantro Sauce

Duet of Grilled Beef and Chicken, Green Peppercorn, Cream and Shiitake Sauce

FISH

Maryland Lump Crab Cake, Dill Butter

Spinach Tortellini, Peppered Shrimp, Fajita Zucchini, Asiago Cream

Grilled Swordfish Steak, Saffron Sauce

Fillet of Salmon on Fennel Ragout and Beurre Rouge Cream

New England Cod, Pine Nut Crust, Pesto Cream

Sauteed Gulf Snapper, Fricassee of White Beans, Cajun Cream

Florida Gulf Shrimp and Jumbo Sea Scallop Brochette
on Herb Fettuccine, Chive Cream

A 19% Service Charge and Applicable State Sales Tax on Food and Beverage will be added. Prices are subject to change without notice.

DESSERTS

Fruit Tart, Raspberry Coulis

Pear and Franzipan Tart

Ginger Carrot Cake

Warm Apple Strudel, Vanilla Sauce

Triple Layer Chocolate Mousse Cake

Tanqueray Parfait, Honied Grapefruit

Seasonal Berries, Creme Anglaise

Key Lime Pie

Hazelnut Torte, Chocolate Sauce

DELUXE DESSERT SELECTIONS

Opera Cake, Chocolate Ladle, Cappuccino Sauce

Dual Creme Gratinée

Duet of Mini Pastries

Raspberry Crème Brulée

Chocolate Crème Brulée

Golden Ingot

Chocolate Decadence

A 19% Service Charge and Applicable State Sales Tax on Food and Beverage will be added. Prices are subject to change without notice.

♦ **FIGURE 4-13.** A la carte pricing is the concept of the catering menu item selection listing from Ritz Carlton Hotels. Reprinted with permission of Ritz Carlton Hotel Philadelphia.

Step 2. Once all of the menu items have been given a minimum price, the main course section of the menu is reviewed. Establish a price range for the main course section, noting the highest and lowest prices. Ideally, the most expensive items should only cost $10 more than the least expensive item, not including seasonal items and daily specials. After the entree prices are set, all the surrounding menu items can be priced. Categories, of course, should be priced so that each section's prices are balanced in relation to the others and make sense to the customer. Having appetizers priced at $3.50 and soups at $1.95 would not seem logical to the customer. If the appetizer price is $3.50, the soup price should be $2.00 or higher to create balance. On the same menu, dessert prices would probably range between $2.50 and $4.00. If too wide a range of prices is offered, the customer will be caught up in a question of expense, trying to decide whether to select a high-priced item or choose a "bargain."

Step 3. The decision should be made as to how to market, or spell out, the actual menu prices. Three methods of decimal placement are commonly used in pricing—the quarter method (0.25, 0.50, 0.75, and 1.00); the nickel-and-dime method (0.05, 0.10, 0.15); and the penny method (0.98 and 0.99). Surveys have shown that most restaurants seem to prefer the quarter method.

Marketing the actual prices requires sensitivity to customer reaction. Customers understand that the restaurant is not a supermarket or department store. A price of $16.98, for example, would be more appropriate for a shirt than a prime rib dinner in a family-style restaurant. Prices should seem appropriate to the customer, just as the price range should.

Step 4. Once the new selling prices have been set, the overall menu should be reviewed to make sure that prices actually match items. The newly constructed prices may be above or below the customer's perceived value. Items that will probably not have heavy sales, for instance, may have been priced too low to make up for possible cost of raw food. On the other hand, an item that has a high food cost, but is expected to be a big seller, can be given a lower selling price because of the volume of sales. Keep in mind that a menu item can always be upgraded to increase quality and food cost (if it has been priced too high), or if it has been price too low, changed to lower the food cost while retaining quality (by decreasing the portion size, for example). However, if the menu copy specifies that an item is of a particular size or quality, management cannot change these specifications unless the menu description is changed as well. Truth in menu applies as much to quick changes in food production policies to compensate for rising food costs as it does to the original menu copy. By planning menu items that can act as reconstitution specials or take advantage of seasonal low prices, management can create a safety valve or cushion against rising food costs on established menu items.

Step 5. The customer's perception of the value of the product should be determined. Due to factors completely beyond a manager's control, customers may have a perceived value for a menu item that is far different from the actual calculated price. Marketing efforts by other companies, social and status pressures, and preconceived viewpoints all influence what a customer believes the item is worth and what the price should be. Imported items, for instance, are usually given higher prices becaue they are considered more desirable. A hamburger does not have the same perceived value as chopped sirloin steak with a wine and mushroom sauce. Strawberries

Romanoff is acceptable at $5.50, while strawberries and cream would rate only $3.50. Here again, conducting a daily survey of items sold can help to determine customer reaction.

Reviewing the reactions of customers is as important as developing the initial menu prices. To determine the reactions of customers, the manager can use three basic forms—the production sheet, the sales mix, and the sales history. These forms are discussed in Chapters 3 and 4. The resulting data are then reviewed and the menu pricing structure changed accordingly. Chapter 4 suggests additional indicators to watch for and various ways to respond to customer reaction. It is important to remember that no pricing structure should be unchangeable. A pricing structure that is designed to be flexible can be adjusted easily to suit production and management.

MAINTAINING PROFITABLE PRICING

The Sales Mix

The sales mix is a representation of the sales of each item on the menu in chart form and is used to determine the average number of entree selections served daily. Using the information from the production sheet and sales history, the foodservice manager charts the popularity and selling trends of each menu selection. By doing so, the manager or operator can see which items are selling poorly, determine when customers prefer certain types of items seasonally, conduct market research, and base future menu plans on concrete evidence of customer likes and dislikes.

The sales mix chart represents items in the major course categories of appetizer, soup, entree, and dessert, and covers a period of up to 30 days. In the sales mix, information from the production sheet and the sales history is consolidated to form a bank of information which can be used to forecast future business and analyze past and present trends.

The sample sales mix in Figure 4-14 represents a 30-day period with six days (Mondays excluded) to each week. For our purposes, a record of a three-week period has been maintained for the entree category while a one-week record has been kept for the dessert, appetizer, and soup sections.

For the first week, the number of veal cordon blue entrees sold fluctuates from Tuesday to Thursday, then levels off and slowly increases through Saturday. On Sunday, sales for this entree drop again, reflecting the overall decrease in covers. In the second week, veal cordon bleu charts a gradual increase of sales throughout the week but does not meet the first week's sales. Compared with the first week, 70 fewer veal dishes were sold. The entire entree count for the second week is down 200 from week number one, with the total veal items representing 35 percent of the decrease. In the third week, 253 veal cordon bleu entrees are sold, only 70 fewer than the first week. Daily third-week sales are stable as well, rising gradually from Tuesday and dropping off only on Sunday as expected.

The decrease in second-week sales for veal cordon bleu may be attributed to a number of different causes. To begin with, overall entree sales were reduced in the second week, which shows that lower sales are not due solely to a dropoff in veal cordon bleu. To determine the actual causes—customer likes and dislikes, the weather, group parties, or the promotion of other menu items, for example—the foodservice manager should consult the sales history for that week.

SALES MIX		
ITEMS: *Appetizer, Soup, Entrée, Dessert*		TIME PERIOD: *18 Days*
COMPILED BY:		

MONTH: October	T 1	W 2	T 3	F 4	S 5	S 6	T	T 8	W 9	T 10	F 11	S 12	S 13	T	T 15	W 16	T 17	F 18	S 19	S 20	T	T W T F S S 22 23 24 25 26 27	T T W 29 30	Average #
Appetizer:																								
1. Oysters	54	47	68	83	156	191	499																	
2. Escargot	32	31	37	49	75	52	276																	
3. Melon & Pros.	25	20	31	37	54	35	202																	
4. Artichoke	29	30	38	52	65	60	274																	
Soup:																								
1. French Onion	45	49	42	54	78	61	329																	
2. Carrot	13	10	8	15	28	20	94																	
3. Cream of Mushroom	32	38	36	46	59	48	259																	
Entree:																								
1. Veal Cordon Bleu	26	20	38	54	66	55	259	21	25	27	35	41	37	186	27	29	32	60	63	42	253			39
2. Tournedos	38	43	52	62	78	66	339	33	38	45	48	56	47	267	32	40	47	54	68	54	295			50
3. Prime Rib	45	40	47	68	80	61	341	42	40	48	55	61	42	288	49	45	54	60	72	58	338			54
4. Scampi Shrimp	12	15	25	30	44	32	158	18	15	20	26	32	30	141	21	18	23	32	39	31	164			26
5. Salmon Steak	8	9	12	18	25	19	91	5	8	11	15	19	20	78	7	6	15	20	24	20	92			15
6. Scallops	21	17	22	31	48	33	172	17	21	23	27	31	26	145	23	32	25	30	35	30	175			27
7. Filet of Scrod	18	22	28	30	27	25	96	17	19	21	26	29	31	143	20	27	25	33	31	28	164			25
8. Lamb Chops	9	12	11	15	18	12	77	8	10	12	16	19	15	80	8	17	13	18	20	14	90			14
9. Chicken Forestiere	4	2	6	7	9	6	34	2	3	5	9	12	8	39	12	10	8	13	9	12	64			7
10. Roast Breast Capon	30	28	41	62	68	54	283	29	32	40	49	61	52	263	31	40	51	54	62	50	288			46
Dessert:																								
1. Mousse	30	24	35	51	108	58	306																	
2. Apple Pie	21	20	29	47	96	51	264																	
3. Hazelnut Torte	41	35	39	48	112	48	323																	
4. Sherbet Bombe	18	25	34	52	78	50	257																	
5. Fresh Fruit	11	21	28	35	47	38	108																	
TOTAL ENTREE:	211	208	282	377	463	363	1850	192	211	252	306	361	308	1630	230	264	293	374	423	339	1923			
TOTAL APPETIZER:	140	128	174	221	350	238	1251																	
TOTAL SOUP:	90	97	86	115	165	129	682																	
TOTAL DESSERT:	123	125	165	233	441	245	1332																	

◆ **FIGURE 4-14.** Sample sales mix chart. The sales mix is a record of the daily sales of each item on the menu.

The sales mix figure, which represents the average number of portions sold in a given period, is used to rate the relative popularity of items in one category. As shown in Figure 4-14, prime rib is the most popular item, with tournedos close behind. Roast breast of capon is third in popularity, with veal, scallops, and scrod in the next three positions. Salmon and lamb are almost equal in popularity, whereas chicken forestiere lags behind with an average of only seven covers per day. Sales mix figures act as a barometer of item popularity, when viewed over a long period of time. Marked customer preferences for a particular item may warrant marketing changes or future promotions for related menu items.

In addition to item popularity, sales mix figures indicate trends in item sales along with losers and leaders, the least and most popular items on the list. Trends in item sales reflect customer reaction toward general types of food items. In our sample chart, the sales mix figures show that beef entrees captured over 70 percent of average sales. Customers are buying more beef than fish items, and their reaction to the poultry category seems to be mixed, since the two poultry items are at opposite ends of the sales mix. The results of customer purchasing preferences show strong sales in beef items, even though there are only three beef items on the menu.

Losers and leaders are the items that lead and trail the sales mix list. In this case, the leaders are prime rib, tournedos, and breast of capon, the losers, lamb chops and chicken forestiere. Salmon steak is a borderline item. It, along with lamb chops, could be developed by special promotions and/or placement on the menu. These items also help to add variety and interest to the menu. Both items are purchased in preportioned cuts that are easy to keep frozen or order in limited quantities. All things considered, salmon and lamb should be kept on the menu and promoted more vigorously for increased sales. Chicken forestiere, however, has had spasmodic sales for a three-week period. Sales have risen only in direct proportion to overall cover counts. Since this item must be produced in quantity and has a limited holding period, it would be advisable to remove it from the menu and replace it with another poultry or beef item. Sales of the capon indicate that customers want poultry, but may have been put off by the name, texture, or presentation of the chicken forestiere.

The relationship of each of the courses to overall sales is also indicated on the sales mix. During the first week, 1,850 entrees were sold. Assuming that each of the customers ordered soup, while 72 percent ordered desserts, a marketing analysis would indicate that soup sales should have increased during the month of October. Since soups generally have a high profit margin, promotions should be stepped up to increase sales. The low percentage of soup sales might also indicate the need to change the soup selection. Carrot soup represents only 15 percent of total soup sales. The item should either be promoted to increase sales or replaced with another selection. Management might also consider incorporating a cycle menu program for soup items to create daily and weekly specials.

An alternative format for the sales mix is shown in Figure 4-15. This 15-day layout provides columns for percentage of total sales, sales mix rank, and percentage of total food cost in addition to a total of item sales. Note that totals for each menu category can be calculated on this format. An example of how to carry out the percentage of sales and sales mix rank is provided at the top of the form as well as additional directions for completing a sales mix. This format can be used as a sample worksheet for calculating a sales mix.

Menu engineering, a sales mix analysis that develops these concepts in depth, was developed by Michael Kasavana of Michigan State University. This program takes the information from the sales mix and assigns it a percentage according to overall sales indicating whether the item should be replaced, repositioned, repriced, or retained. Items placed within the percentage framework are characterized as "plowhorses," "stars," "dogs," and "puzzles," a refinement on "losers" and "leaders." The placement of menu items in these categories can then be graphed by management for analysis.

REEVALUATING PRICES

In any foodservice organization, costs and profits should receive constant review, whether on a daily, weekly, or monthly basis. Properly prepared and executed sales mixes, production sheets, and sales histories are essential for these reviews. The quality of the initial cost calculations and the updated cost figures will determine the quality of the review. When a review indicates that any cost percentage is higher than it should be, it is time to reevaluate menu prices.

SALES MIX

(a) For each section of your menu, rank items by sales.

(b) Enter the items you have selected as your final menu listing under the appropriate categories below.

(c) Record the daily number of sales for each item for a two-week period.

(d) Enter the total sales for each item.

(e) Calculate the percentage of total sales for each menu category (course heading) that the total sales for each item represents.

    Total item sales
    + Total course heading sales
    = % Total sales

(f) Rank each menu item according to percentage of sales in each category, beginning with "1" for the highest seller.

The result of the sales mix ranking in each menu category indicates which menu items are producing the most sales.

(g) Enter the food cost percent you have assigned to each item in the column "Food Cost %". Determine which menu items in each menu category are the most profitable. An item that generates a large sales volume and has a low food-cost percentage results in a high profit margin.

By consistently reviewing your sales mix on a regular basis, you can keep abreast of fluctuating sales and the position of high-profit menu items in the sales mix ranking.

SALES MIX

EXAMPLE:	From: 1/1/91						To: 1/15/91									Total Item Sales	% Total Sales	Sales Mix Rank	Food Cost %
APPETIZERS	1	2	3	4	5	6	7	8	9	10	11	12	13	14	15				
Nachos	6	4	7	5	3	8	4	6	8	7	5	3	6	4	5	81	61%	1	
Chicken Wings	4	3	4	5	2	4	1	3	5	4	3	3	4	2	4	51	39%	2	
													Total course heading sales			132	100%		

YOUR OPERATION:	From:	To:		Total Item Sales	% Total Sales	Sales Mix Rank	Food Cost %
APPETIZERS							

Total course heading sales — 100%

SOUPS

100%

SALADS

100%

ENTREES

100%

◆ **FIGURE 4-15.** Sales mix worksheet. This format offers a 15-day overview of item sales. Percentage of total sales for each item along with overall food cost rank can be determined using this format. Reprinted with permission. Copyright 1991 *Menu for Profit*.

SALES MIX

VEGETABLES (If sold separately)	From:						To:										Total Item Sales	% Total Sales	Sales Mix Rank	Food Cost %
								Total course heading sales										100%		

STARCHES (If sold separately)																				
																		100%		

DESSERTS																				
																		100%		

♦ **FIGURE 4-15.** *(Continued)*

The decision to change prices depends on a great many considerations. For instance, how many menu items are actually involved in the cost percentage changes? How are these items reflected in the sales mix? Are these items losers or leaders? If just a few items are involved and if these items are not high on the sales mix, management may choose to change the portion sizes slightly to absorb some of the food cost increases rather than change the price. If, however, a substantial number of popular items show food cost percentage increases, the following points must be considered before a decision is made to change prices.

1. Can these items be promoted at current prices to increase sales so that volume purchase prices will lower the overall food cost?
2. Can the promotion of lower-cost menu items reduced sales of higher-cost items?
3. Can the portion size or the presentation of the item be changed without affecting the customer's perceived value of the item?

Most important is the customer's ability to accept the price change. With this in mind, the manager should consider the perceived value of the item in comparison to the projected menu price increase. If customers do not think that an item has greater value than they are paying for it, they will not tolerate a price increase. In this case, the item must be upgraded in portion size or product quality or dropped from the menu.

The manager's next consideration would be the timing of the last price change. Prices should be changed at regularly scheduled intervals. All prices should be changed at the same time, even if some items are given a price "cushion," or lower food cost percentage, in order to balance the overall pricing structure. The cushion will be absorbed later by increased costs on higher priced items.

A final consideration regarding price changes is whether the menu will need to be reprinted. Changing individual items by hand on the menu is unprofessional and will create a sloppy appearance. Printing new menus, however, is expensive. If menu prices must be changed without reprinting, consider the following mechanical techniques:

1. Use laminated pages or panels, to which special ink adheres.
2. Have the menu printed without prices in the first place, and order a large supply. Using a laser printer and a word processor, change the prices as needed.
3. Design sections into the menu layout for items that are subject to price change. Using a laser printer, replace these sections as necessary.

Whichever method is used to change prices, professionalism and customer acceptance should be the primary considerations.

◆ PART II ◆

Developing
a Menu
Program

CHAPTER 5

**Menu
Item
Selection**

CRITERIA FOR MENU SUCCESS

Determining customer need and market feasibility are only the initial steps in planning a successful menu program. Before the design, color, and format of the printed menu are considered, a series of decisions must be made about the contents of the menu. In addition to choosing the right menu items, standards of preparation, production, and presentation must be established to create consistent quality controls for all menu items.

In the beginning stages of menu planning, five questions need to be answered:

1. What is the style and/or theme of the restaurant?
2. What is the established cuisine?
3. What are the needs of the customer?
4. What is the restaurant staff capable of producing?
5. What menu items will be marketable and profitable?

Service Style

Service style is the format in which food is served to the customer. Fast food, family style, cafeteria style, French service, Russian service, buffet service, and American service are the different styles of service commonly used in foodservice operations.

The choice of service has a great deal to do with the planning of the menu. The cuisine of the restaurant often determines the style of service. Some cuisines require combinations of buffet and table service, tableside preparation (French service), or Russian service. However, almost every cuisine can be adapted to a particular style or combination of styles. American restaurants in the 1990s have capitalized on casual-style themes and service. Service for the casual-style restaurant operation is based on American service with food items plated in the kitchen. While the table service guidelines of: serve to the right, clear to the left, beverages always served to the right are the ideal, table configurations combined with casual formats and atmospheres often result in "service as it happens . . . just get it to the table."

The needs of the customer are also important to consider. What will the customer feel comfortable with? If buffet style is part of your service plan, is there enough floor space to accommodate this style of service comfortably or will it be hard to maneuver between tables and other guests with plates of food in hand? French service may be too showy or sophisticated. Will your customers feel at ease or would they prefer the simplicity of American service? The selection of a service style should also be based on the ability of the available staff to perform the service and/or management's ability to train them. If finding quality service help is a problem, then establishing an elaborate service style for the restaurant would not be advisable.

Profit also influences the serving style. French service is labor intensive and requires more equipment, more service time, and results in slower table turnover rates, all of which affect the profit margin if the increased costs cannot be covered in menu prices. Therefore, you must consider the following:

♦ Will the customer perceive enough value in your menu selections and service to pay the menu prices?

◆ Will the menu prices be high enough to cover the costs of both food and labor?

Self-service is suited to a casual style of dining, lower menu prices, lower labor costs, and higher table turnover rates. Russian and American service both complement full-service restaurants and fall in between the demands of French service and unstructured self-service. Both of these styles have equipment needs, but servers do not have to be highly skilled in tableside presentation. Regardless of the service style, customer turnover rates are controlled by the ability of the kitchen and the dining room staff to prepare food items and deliver service to the kitchen. With the shortages in foodservice labor that will continue to affect the foodservice industry at all levels, this consideration is often primary in making the appropriate choice of service style.

Cuisine

Cuisine refers to the style or manner in which food is prepared. As a foodservice industry term, it is usually associated with a specific nationality, such as Italian, Chinese, or German. Cuisine refers to the cooking styles of certain regional areas (Southern, French Provençal, or New England) and to such cultural groups as Creole, Shaker, and Pennsylvania Dutch. Cuisine also applies to general cooking styles such as classical or nouvelle.

Although individual foodservice operators have a fixed or predetermined idea about the kind of cuisine they would like to offer, the market survey will help to identify what would be the most successful. It is important to analyze existing restaurants in the area and determine which cuisines are currently successful. Very often, the style of cuisine that the foodservice operator prefers can be tailored to suit the community, if the need is there. For example, a prospective restauranteur wants to open a family-style Italian restaurant, offering a full-service menu with a selection of Italian foods, rather than fast-food items such as pizza or Italian sandwiches. The market survey indicates that a sizable number of small fast-food restaurants are located in the area. There are also two family-style Italian restaurants just above the fast-food level, offering basic Italian menu items. All of the restaurants are doing reasonably well, pointing to a high consumer response to Italian food. The results also indicate a need for a full-service restaurant offering table service and a semi a la carte menu. The prospective restauranteur now needs to decide whether to alter plans enough to develop a full-service restaurant or try another location and/or cuisine.

Customer Needs

Customer needs relate directly to the dietary needs, preferences, and tastes of the individual. An example of identifying customer needs is seen in the response of the restaurant industry, as a whole, to increased awareness of health and nutrition on the part of the general public, combined with an ongoing concern with cholesterol, sodium, and caloric counts. Restaurant menus are being developed that specifically

APPETIZERS

GULF SHRIMP COCKTAIL			7.25
LOBSTER STEW			
Cup 5.75		Bowl	11.50
SAIL LOFT CLAM CHOWDER			
Cup 4.25		Bowl	8.50
STEAMED NATIVE CLAMS			
Small 7.75		Large	15.50
MUSSELS STEAMED IN WINE			7.25
FRENCH ONION SOUP			
Baked in a cheese crusted crock.			
Small 4.25		Large	6.50

BLACKBOARD SPECIALS

AUGUST 4, 1997

SAIL LOFT AWARD WINNING FISH CHOWDER
Cup 3.50 Bowl 6.50

SPINNEY CREEK OYSTERS 8.95
Shucked to order, on the half shell

ESCARGOTS A LA BOURGUIGNONE 6.95
Snails baked in a traditional brandy, garlic &
butter sauce

SPINNEY CREEK OYSTER STEW
Cup 5.50 Bowl 10.50

MAINE SHRIMP & PESTO 15.50
with white wine on a bed of linguine

FRESH LOCAL HALIBUT 17.95
Baked in a leek & mushroom chardonnay sauce

BROILED FRESH ATLANTIC SALMON 17.95
With béarnaise sauce

BROILED FRESH TUNA 17.95
with an orange glaze

LOW FAT CHOICE

LEMON BROILED HALIBUT 17.95
served on a bed of crisp greens
with fresh raspberry vinaigrette

DINNER ENTREES

LINGUINE WITH MAINE SHRIMP AND LOCAL SCALLOPS/17.00	
In a roasted tomato, garlic, basil, and white wine sauce.	15.50
SEAFOOD PASTA SALAD	15.50
Local seafood and assorted pasta in a creamy garlic dressing	
MAINE SHRIMP CAESAR SALAD	15.50
Local shrimp sauteed in olive oil served on a bed of classic Caesar Salad.	
CRABMEAT CAESAR SALAD	16.25
Fresh local crabmeat on a bed of classic Caesar Salad.	
PENOBSCOT BAY SCALLOPS	17.00
Broiled or baked with white wine and garlic butter, or fried and served with tartar sauce.	
FISHERMAN'S PLATTER	18.50
A broiled or fried platter of clams, scallops, shrimp, fish and steamed mussels. We suggest the clams be fried or steamed.	
LOBSTERMAN'S PLATTER	24.50
The above with one half of a steamed lobster in the shell.	
BAKED STUFFED SHRIMP	20.50
Jumbo shrimp baked with our own bread crumb dressing.	
STEAK, LOBSTER AND MAINE SHRIMP	24.50
Sliced filet mignon with one half of a steamed lobster in the shell and Maine shrimp baked in white wine and garlic butter.	
BAKED LOCAL HADDOCK	18.50
With crabmeat stuffing.	
LOCAL MAINE SHRIMP	15.50
Baked with white wine and garlic butter or fried and served with tartar sauce.	
FRIED NATIVE CLAMS	16.75
We shuck our own.	
ROAST DUCKLING WITH MOLLYS BLUEBERRY CHUTNEY SAUCE	17.95
Crisp half duckling served with wild rice pilaf.	
BROILED CHOPPED BEEF STEAK	10.95
With sauteed onions and mushrooms.	
SLICED FILET OF BEEF	22.00
Slices of tenderloin of beef sauteed in butter.	
FILET MIGNON	23.00
A thick cut of steer tenderloin broiled over the open flame. Served with Bearnaise sauce.	
BROILED CHICKEN	10.95
Half a broiler, partially boned.	
MEDALLIONS OF CHICKEN	15.95
Sauteed in butter with local crabmeat and Bearnaise sauce.	

LOBSTERS

SAIL LOFT SHORE DINNERS	
First, clam chowder Next, steamed native clams or mussels and then, a steamed lobster.	
1 pound lobster	26.50
1½ pound lobster	33.50
2 pound lobster	40.25
BAKED STUFFED LOBSTER	
This lobster is opened and filled with a special Sail Loft filling and more lobster pieces and baked until tender.	
1½ pound lobster	26.75
2 pound lobster	35.00
STEAMED LIVE LOBSTERS	
Cooked to order.	
1 pound lobster	15.75
1¼ pound lobster	19.25
1½ pound lobster	22.75
2 pound lobster	29.50
Larger sizes available.	
LOBSTER SAUTE	29.50
Large chunks of lobster removed from the shell and sauteed in butter.	
LOBSTER NEWBURG	22.00
We saute lobster meat in butter and sherry, then simmer in our Sail Loft Newburg Sauce and serve in a casserole with toast points.	
LOBSTER SALAD	27.50
A salad of freshly opened lobster meat served in the shell.	

Live lobsters are brought directly to our salt water tanks. From there they are cooked to order or for lobster meat, prepared each day.

The SAIL LOFT is open year round for Lunch and Dinner seven days a week. Sunday Brunch also. We would be pleased to host your next special occasion. Most of our menu items can be prepared and packaged to take with you. As part of ROCKPORT MARINE we offer complete wooden boat storage, repair, restoration and building services.

Taylor Allen   Lucia and Barry Thompson
Boatyard, 207-236-9651   Restaurant, 207-236-2330

We will be happy to prepare any dish to accommodate your dietary restrictions.

♦ **FIGURE 5-1.** Menu items with nutritional content identification. Many menus offer the customers a breakdown of caloric, fat, sodium, and cholesterol content of the ingredients of certain menu items. Reprinted with permission of the Sail Loft, Rockport, Maine.

identify items that are "health" and "wellness" related, as in the menu in Figure 5-1. These items are often listed along with appropriate nutritional information on the menu, as in Figure 5-2. Fast-food giants like McDonald's have responded to customer concerns with a fresh salad selection. Advertising and merchandising often now include the words "natural" as well as "fresh" to distinguish between those food items containing artifical colors are preservatives and those items considered uncontaminated.

Customer needs are also evident in the demographic makeup of a community. If students, tourists, or working women are identified as a large socio-economic group within the area, the needs of these customers should be considered in planning menus. Students will need large portions at low prices. Tourists may want to try regional cuisine items, and working women will need "home replacement" food items packaged for take-out or inexpensive family meal alternatives. Menus from

GRILLED TUNA SANDWICH

NUTRITIONAL MODIFICATION:
USE LOW CHOLESTEROL MAYO/MUSTARD DRESSING
> TOTAL CALORIES: 358   CHOLESTEROL: 83 MG
> FAT: 7.4 GRAMS   SODIUM: 872 MG
> % CALORIES FROM FAT: 18 %

♦ **FIGURE 5-2.** This presentation specification offers a color photograph of the finished and plated menu item along with the nutritional content, ingredients, quantity, cost, and description of plate setup.

your direct competition are a good guideline to local customer needs and the general tastes of the community. The appetizer, salad, main course, and dessert selection from the competition's menus are a good indication of what the community generally finds acceptable. If four out of five menus surveyed offer stuffed mushrooms as an appetizer, this indicates that stuffed mushrooms should be a popular menu item. If all five restaurants provide a wide variety of desserts, it can be assumed that dessert selections are popular with local customers. What you cannot see, however, is how well these items are selling. The only way to be sure of customer needs is to test them for popularity.

Using competitors' menus as a guideline does not mean that they should be duplicated. Although it makes sense to adapt what seems to be successful into your own menu, customers are looking for variety.

Once the menu is in operation, customer needs can be determined more accurately from the results of the sales mix. Items that are not selling over a period of time should be analyzed. Customers may not be interested because they are bored with the duplication between one restaurant and another. Many casual-style restaurants today are offering menus that are very similar in content. When moving from one to another, it is often difficult to find different menu items, especially on the appetizer and dessert sections.

Lack of customer interest in menu items can be the result of inadequate marketing. Poor placement on the menu and misleading accent and descriptive copy may cause the customer to miss the item in the selection process or be unable to understand what the item is. It is important to review these possibilities before removing an item from the menu. Asking for customer comments is also a good way to evaluate customer needs. Place comment cards in accessible locations, such as tent cards or in check presentation folders, to give customers a chance to note their preference—likes and dislikes.

Equipment Capacity

Assessing the ability of your operational staff of produce menu items is important to the success of the menu program. Planning to serve menu items that staff and/or equipment are not capable of producing can cause the menu program to fail. Equipment is the most important consideration in evaluating the production potential of an operation. In a new operation, the menu should be planned before equipment is purchased, so that the menu and the equipment can be coordinated for maximum production capability and efficiency. In a building where equipment already exists or in a situation in which the menu program needs a complete overhauling, the available equipment should be reviewed before the menu is planned.

First, consider the various cooking methods available or called for. Does the kitchen have or need ranges, broilers, ovens, or steamers, for example, to produce the menu? The second consideration for equipment is cooking load. Cooking load is the total amount of food that can be cooked either in the kitchen or on any one piece of equipment at a given time. When planning the menu or determining what equipment to buy, establish what the cooking load capacity will be at peak times of service. A general menu that can be produced easily when the kitchen is slow may not necessarily be easy to produce when the restaurant is busy. Take into account what each menu item requires in terms of preparation so that production and service will run smoothly at peak times. Match the methods of preparation to the equipment to make sure that methods and machines are equally distributed, producing a balanced production line. The work page in Figure 5-3 can help to determine equipment needs for a menu program.

Staffing Capacity

Staff is an important operational consideration. How well the staff are trained and how capable they are of producing involved, intricate, and highly technical food presentations will influence the success of the menu program. Plan for staffing problems such as absenteeism and staff shortages. Assume that the menu will have to be produced under pressure.

**RESEARCH FACILITIES &
EQUIPMENT AVAILABLE**

Storage area — square footage:

Preparation area — square footage:

Dining/service area — square footage:

— number of seats:

Are these facilities adequate for the style of dining that you offer?

Kitchen equipment in place:

_____	_____
_____	_____
_____	_____
_____	_____
_____	_____
_____	_____
_____	_____
_____	_____

Review the type of cuisine you have selected. List any additional kitchen equipment you need to purchase in order to offer this type of cuisine.

_____	_____
_____	_____
_____	_____
_____	_____
_____	_____
_____	_____
_____	_____
_____	_____

◆ **FIGURE 5-3.** Worksheet for facilities and equipment assessment. Reprinted with permission. Copyright 1991 *Menu for Profit.*

There are critical shortages of qualified foodservice labor in the United States today. The industry, as a whole, is short approximately 200,000 workers every year (National Restaurant Association). Most operators open their doors every day knowing that they are understaffed in either the kitchen, dining room, or both. The most critical area of shortage is for mid-level staff that have had training in food preparation or table service.

Menu items should be selected for which quality, preportioned, and partially preprepared food items can be obtained from area purveyors and distributors. Food manufacturing companies are continuously developing products to answer this need. This does not mean that every menu should be made up of very simply prepared items. Menus should, however, be balanced in preparation and production techniques, so that complicated, involved preparations are offset by simpler items that do not require highly skilled help. Let the trained personnel produce high-profit specialties while the regular staff produces the balance of the menu.

THE MENU REPERTORY

Menu planning, like every phase of the foodservice industry, requires organization. The menu planner searches for food items that will satisfy the needs of the customer, restaurant, and profit margin and suit the style and cuisine of the overall operation. Once found, these items should be put into some form of organization so that they can be used to the maximum benefit of the menu program. The menu repertory is a file system that identifies all the food items a kitchen can produce, provides a cross reference for recipe and costing information, and supplies easy guidelines for using and matching the different items.

Every item listed in the menu repertory is assigned an index number and backed up with specifications, a standard recipe card, and a standard cost card. Accompanying this main list are the recipe card and cost cards for the stocks, sauces, doughs, and other major ingredients used in the recipes. The purpose of the file is to provide the executive chef and foodservice manager with all of the information they need regarding the menu. The file contains a categorized list of recipes, each of which has been tested for production consistency and given an updated cost card that can be accurately priced at the determined food cost percentage.

When the general outlines for the selection of menu items have been established (according to cuisine, customer needs, dietary trends, production requirements, and desired average check), the process of menu item selection can begin. This process is illustrated in Figure 5-4.

Selection Criteria

In the first phase of menu selection, all of the items in a given food category that meet the general requirements are considered. In the second phase, any items that present production, cost, truth in menu, or other problems are eliminated.

Equipment capability is a major concern in selecting menu items. As discussed previously, the menu planner should consider the types of equipment and the cooking load capacity for each piece of equipment. Recipes will be selected according to the ability of the kitchen to produce the finished item at peak periods. Each item is evaluated for preparation and holding problems. Foods that hold up well for periods of time under refrigeration and reheating are wise choices. Those foods that have a tendency to break down within a short period of time or change color would not be suitable.

A menu selection that is time consuming to prepare or has highly technical production steps should be eliminated unless production capability is guaranteed

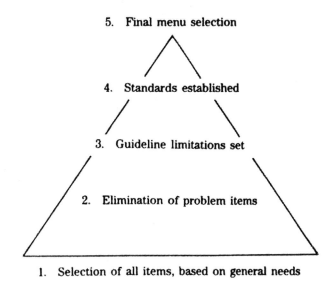

5. Final menu selection

4. Standards established

3. Guideline limitations set

2. Elimination of problem items

1. Selection of all items, based on general needs

♦ **FIGURE 5-4.** The steps of elimination that take place in developing the menu repertory.

or special considerations exist. As discussed earlier, the staff's ability to produce a menu item is a paramount consideration in the overall selection process. The suggestion to develop menu items around the availability of preprepared products does not mean that quality should be sacrificed. On the contrary, by choosing quality products as a base for production, the consistency of the finished item is more controllable. By increasing a portion of the food cost and decreasing labor cost, the overall differences between using high-quality precut meats that have been vacuum packaged to preserve freshness, and buying a side of beef that must be broken down into usable portions and waste, are balanced out. The comparison shown in Figure 5-5 indicates the advantages of choosing the preportioned, partially prepared item over the raw bulk product.

Cost is the next consideration. Choosing fish-based soups made with shellfish, for example, can be costly. A she-crab soup or lobster bisque looks good on the menu but could be expensive to produce if lobster is not on the regular menu, to be reconstituted into the bisque. She-crab soup demands a fresh supply of crab and crab roe; a good shrimp bisque is made with shells and pieces of shrimp. By offering a simple seafood bisque, the luxury can be offered without promising the customer a specific type of shellfish or causing problems for the restaurant if the item is not available or is running at a very high cost. If a variety of fish items is offered on the menu, the kitchen can produce a bisque made of whitefish, shrimp, scallops, crab, or any item that is currently being offered elsewhere on the menu. A look at the seafood soups offered at the Sail Loft restaurant in Figure 5-6 shows a variety of recipes including fish chowder, clam chowder, oyster stew, and lobster stew. As this restaurant specializes in Maine seafood, maximum use is made of trimmings and extra portions of seafood and shellfish in this selection.

Truth in menu is a regulation similar to the truth in advertising law. Customers

BEEF ITEM	FINISHED	
#103 Rib of Beef: 35 lb	#112 Rib Eye Roll (net weight): 11 lb	
Cost per lb: $2.62	Cost per lb: $4.95	
Purchase price: $91.70	**Purchase price: $54.45**	
Raw waste after trimming: 24 lb	No trimming	
Cost of waste:	$62.88	Cost of waste: $0.00
Cost of labor to butcher:	5.00	
Total cost waste:	67.88	
By products from waste		
1lb short ribs:	$2.45	
4lb beef cubes:	10.40	
1½lb ground beef	2.77	
	15.62	
Value of waste:	15.62	
Actual cost waste:	**$52.26**	**Actual cost waste:** $0.00

Actual cost product as used (per pound):

#103 Rib of Beef:	**$7.00**	#112 Rib Eye Roll:	**$4.95**

The actual cash loss on this beef purchase for the #103 Rib of Beef was $52.26. If the #112 Rib Eye Roll had been selected initially the cost of the finished, quality product would have been slightly more ($54.45) than the actual cost of the waste of the #103 ($52.26).

BEEF ITEM	FINISHED	
#189 Full Tenderloin, Fat On, purchase weight: 9 lb	#189A Full Tenderloin, defatted, purchase weight: 5 lb	
Cost per lb: $7.15	Cost per lb: $9.65	
Purchase price: $64.35	**Purchase price: $48.25**	
Raw waste after trimming: 4 lb	No trimming	
Cost of waste:	$28.60	Cost of waste: $0.00
Value of waste:	$0.00	
Actual cost waste:	**$28.60**	**Actual cost waste:** $0.00

Actual cost product as used (per pound):

		#189A Full Tenderloin	
#189 Full Tenderloin, Fat On:	**$9.65**	Defatted:	**$9.65**

The actual cash loss on this beef purchase for the #189 Full Tenderloin Fat On was $28.60. If the #189A Full Tenderloin Defatted had been selected initially the cost of the finished quality item would have been $48.25, considerably less than the as-purchased price of $64.35 for the #189 Full Tenderloin.

♦ **FIGURE 5-5.**   Cost yield comparisons of bulk and trimmed, packaged beef. (*Source:* George L. Wells Meat Company, Philadelphia, PA.)

must receive the product advertised on the menu. Large gulf shrimp are not prawns, spring lamb is not mutton, cornish game hens are different from broilers, and hot fudge is not warmed up chocolate sauce. Compliance with truth in menu is best handled at this stage of menu development. Any item that cannot be changed or presented to the public truthfully should be eliminated. Using Figure 5-1 again as an example, native clams indicate "area New England clams," local crabmeat means Maine coast crabs, Maine shrimp is from Maine coastal waters—not Georgia, spinney creek oysters must originate from that area, and a hydroponic tomato is specially grown, not the standard bulk variety.

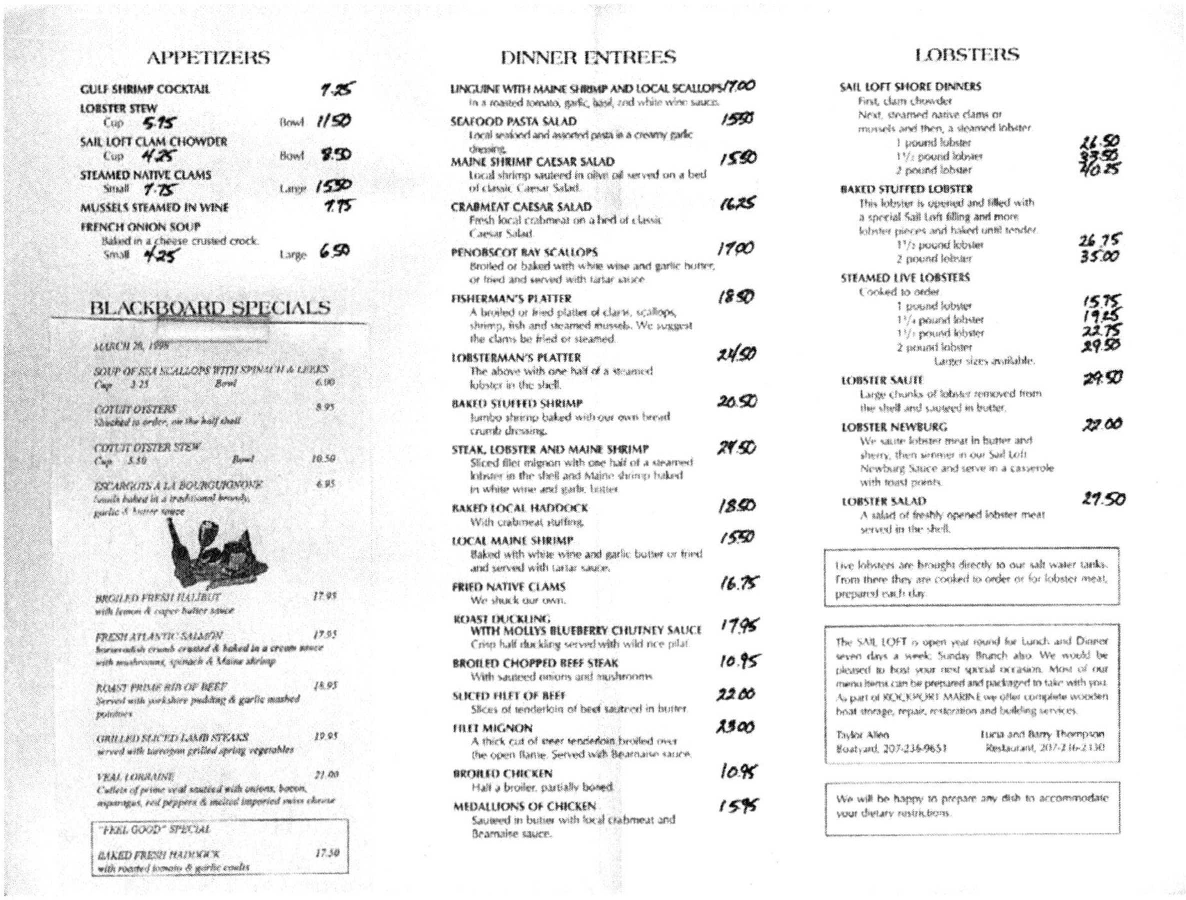

APPETIZERS

GULF SHRIMP COCKTAIL — 7.75

LOBSTER STEW
Cup 5.75 — Bowl 11.50

SAIL LOFT CLAM CHOWDER
Cup 4.25 — Bowl 8.50

STEAMED NATIVE CLAMS
Small 7.75 — Large 15.50

MUSSELS STEAMED IN WINE — 7.75

FRENCH ONION SOUP
Baked in a cheese crusted crock.
Small 4.25 — Large 6.50

BLACKBOARD SPECIALS

MARCH 28, 1998

SOUP OF SEA SCALLOPS WITH SPINACH & LEEKS
Cup 3.25 — Bowl 6.00

COTUIT OYSTERS — 8.95
Shucked to order, on the half shell

COTUIT OYSTER STEW
Cup 5.50 — Bowl 10.50

ESCARGOTS A LA BOURGUIGNONE — 6.95
Snails baked in a traditional brandy,
garlic & butter sauce

BROILED FRESH HALIBUT — 17.95
with lemon & caper butter sauce

FRESH ATLANTIC SALMON — 17.95
horseradish crumb crusted & baked in a cream sauce
with mushrooms, spinach & Maine shrimp

ROAST PRIME RIB OF BEEF — 18.95
Served with yorkshire pudding & garlic mashed
potatoes

GRILLED SLICED LAMB STEAKS — 19.95
served with tarragon grilled spring vegetables

VEAL LORRAINE — 21.00
Cutlets of prime veal sautéed with onions, bacon,
asparagus, red peppers & melted imported swiss cheese

"FEEL GOOD" SPECIAL

BAKED FRESH HADDOCK — 17.50
with roasted tomato & garlic coulis

DINNER ENTREES

LINGUINE WITH MAINE SHRIMP AND LOCAL SCALLOPS — 17.00
In a roasted tomato, garlic, basil, and white wine sauce.

SEAFOOD PASTA SALAD — 15.50
Local seafood and assorted pasta in a creamy garlic
dressing.

MAINE SHRIMP CAESAR SALAD — 15.50
Local shrimp sautéed in olive oil served on a bed
of classic Caesar Salad.

CRABMEAT CAESAR SALAD — 16.25
Fresh local crabmeat on a bed of classic
Caesar Salad.

PENOBSCOT BAY SCALLOPS — 17.00
Broiled or baked with white wine and garlic butter,
or fried and served with tartar sauce.

FISHERMAN'S PLATTER — 18.50
A broiled or fried platter of clams, scallops,
shrimp, fish and steamed mussels. We suggest
the clams be fried or steamed.

LOBSTERMAN'S PLATTER — 24.50
The above with one half of a steamed
lobster in the shell.

BAKED STUFFED SHRIMP — 20.50
Jumbo shrimp baked with our own bread
crumb dressing.

STEAK, LOBSTER AND MAINE SHRIMP — 24.50
Sliced filet mignon with one half of a steamed
lobster in the shell and Maine shrimp baked
in white wine and garlic butter

BAKED LOCAL HADDOCK — 18.50
With crabmeat stuffing.

LOCAL MAINE SHRIMP — 15.50
Baked with white wine and garlic butter or fried
and served with tartar sauce.

FRIED NATIVE CLAMS — 16.75
We shuck our own.

ROAST DUCKLING
WITH MOLLYS BLUEBERRY CHUTNEY SAUCE — 17.95
Crisp half duckling served with wild rice pilaf.

BROILED CHOPPED BEEF STEAK — 10.95
With sautéed onions and mushrooms.

SLICED FILET OF BEEF — 22.00
Slices of tenderloin of beef sautéed in butter.

FILET MIGNON — 23.00
A thick cut of inner tenderloin, broiled over
the open flame. Served with Bearnaise sauce.

BROILED CHICKEN — 10.95
Half a broiler, partially boned.

MEDALLIONS OF CHICKEN — 15.95
Sautéed in butter with local crabmeat and
Bearnaise sauce.

LOBSTERS

SAIL LOFT SHORE DINNERS
First, clam chowder.
Next, steamed native clams or
mussels and then, a steamed lobster
1 pound lobster — 26.50
1½ pound lobster — 33.50
2 pound lobster — 40.25

BAKED STUFFED LOBSTER
This lobster is opened and filled with
a special Sail Loft filling and more
lobster pieces and baked until tender.
1½ pound lobster — 26.75
2 pound lobster — 35.00

STEAMED LIVE LOBSTERS
Cooked to order
1 pound lobster — 15.75
1¼ pound lobster — 19.25
1½ pound lobster — 22.75
2 pound lobster — 29.50
Larger sizes available.

LOBSTER SAUTE — 29.50
Large chunks of lobster removed from
the shell and sautéed in butter.

LOBSTER NEWBURG — 22.00
We sauté lobster meat in butter and
sherry, then serve it in our Sail Loft
Newburg Sauce and serve in a casserole
with toast points.

LOBSTER SALAD — 27.50
A salad of freshly opened lobster meat
served in the shell.

Live lobsters are brought directly to our salt water tanks.
From there they are cooked to order or for lobster meat,
prepared each day.

The SAIL LOFT is open year round for Lunch and Dinner
seven days a week. Sunday Brunch also. We would be
pleased to host your next special occasion. Most of our
menu items can be prepared and packaged to take with you.
As part of ROCKPORT MARINE we offer complete wooden
boat storage, repair, restoration and building services.

Taylor Allen — Lucia and Barry Thompson
Boatyard, 207-236-9651 — Restaurant, 207-236-2330

We will be happy to prepare any dish to accommodate
your dietary restrictions.

◆ **FIGURE 5-6.** A balanced variety of soup offerings is presented in this menu from the Sail Loft Restaurant. Reprinted with permission of the Sail Loft, Rockport, Maine.

Final Menu Selection

In the third phase of the menu selection, guidelines are established as to the number of items that will be included on the menu. There should always be at least three times as many items in the menu repertory as are offered in the menu. If four soups are to be offered, for example, the soup section of the menu repertory should include twelve different types of soup. The soup section of the menu repertory is then further classified according to consistency, color, and method of preparation, among other characteristics, to ensure that a balanced variety of soups will be available for substitution as seen in Figure 5-6 and outlined in Figure 5-7. The other categories of the menu repertory (meat, fish, and so on) are also classified in this manner, according to the appropriate characteristics of food items and product.

Using the section of the menu repertory in Figure 5-7, the chef will be able to prepare soups to take advantage of seasonal foods, reconstituted leftover items, or simply provide customers with a balanced variety of colors, textures, and temperatures.

The work pages in Figure 5-8 can be used to develop a menu inventory.

Clear	Cream	Chowder	Vegetable-based
Consomme Juliene	Vichyssoise	New England Clam	Gazpacho
Beef Madriline	Cream of Carrot	Oyster Stew	Minestrone
Mock Turtle	Cream of Spinach	Corn Chowder	Gumbo (Beef, Poultry, Seafood)

◆ **FIGURE 5-7.** The soup section of a menu repertory as classified in four categories according to consistency, color and method of preparation.

MENU INVENTORY 1, APPETIZERS TO SANDWICHES

List as many menu items as you can think of that can be produced from your kitchen. Use the left side of each line. Do not try to evaluate the items at this time. You may use the right side of the line for comments later.

Example: Shrimp cocktail   _(Comments)_

APPETIZERS
COLD   HOT

SOUPS

SALADS

SANDWICHES

MENU INVENTORY 2, ENTREES

MEATS
BEEF   PORK

FISH   POULTRY

VEAL   OTHER

◆ **FIGURE 5-8.** Work pages provide the outline for listing an inventory of all of the menu items that a foodservice operation is capable of producing. Reprinted with permission. Copyright 1991 _Menu for Profit._

MENU INVENTORY 3, VEGETABLES

VEGETABLES

_____

_____

_____

_____

_____

_____

_____

_____

_____

_____

_____

_____

STARCHES

POTATO   RICE

_____   _____

_____   _____

_____   _____

_____   _____

_____   _____

_____   _____

PASTA   OTHER

_____   _____

_____   _____

_____   _____

_____   _____

_____   _____

_____   _____

_____   _____

_____   _____

MENU INVENTORY 4, DESSERTS/BEVERAGES

DESSERTS

COLD   HOT

_____   _____

_____   _____

_____   _____

_____   _____

_____   _____

_____   _____

_____   _____

_____   _____

_____   _____

_____   _____

_____   _____

_____   _____

_____   _____

_____   _____

BEVERAGES

HOT   COLD

_____   _____

_____   _____

_____   _____

_____   _____

_____   _____

_____   _____

_____   _____

_____   _____

◆ **FIGURE 5-8.** *(Continued)*

In the fourth phase of menu selection, standards are set for each item in the menu repertory: specifications, a standard recipe card, and a standard cost card. Each item in the menu repertory is tested and retested until a recipe is produced that will consistently yield a certain number of portions of the recipe quality. Next, a standard cost card is prepared for each recipe. The cost card is an accurate statement of all the ingredients, costs, individual portions costs, and the desired food cost percentage from which to calculate the selling price. Cost cards can be updated quickly against current purchasing records. Computer software programs allow purchasing prices to be immediately reflected in the costing process. A specification is made up on any item not currently included in the general purchasing list.

As the selection of menu items goes into the fifth and final stage, a further refinement should take place. Entree selections are chosen first, on the basis of general cuisine, number of entrees to be offered, types of entrees (fish, beef, lamb, veal, for example), and method of preparation (roasting, broiling, grilling, and so on). Decisions are also based on the available equipment and its cooking load

	Glazed Carrots	Broccoli	Mushrooms, sautéed	Green Beans Amandine	Asparagus	Leeks, au gratin	Spinach Souffle	Brussels Sprouts	Zucchini with Tomato and Onion	Duchess Potatoes	Chateau Potatoes	Rice Pilaf
Salmon Steak												
Filet of Sole												
Swordfish												
Prime Rib												
Braised Beef												
Sirloin Steak												
Sautéed Veal												
Veal Breast												
Lamb Chops												
Poached Chicken												
Chicken Breast												
Duck Breast												
Duck a l'Orange												

♦ **FIGURE 5-9.** Be referring to a menu suitability chart such as this, a useful checklist of all entrees and surrounding items included in the repertory, the chef can combine or substitute different items to take advantage of daily specials or seasonal foods.

capabilities. After the entrees are chosen, the remaining items are selected by category and then balanced against entrees. Vegetables, for instance, were previously chosen as a group in phase three and balanced for contrasting texture and color. A selection of creamed spinach, sauteed zucchini and onion, creamed corn, and whipped butternut squash would have provided variety of colors, but the textures of all four items are creamy or runny. French green beans, kernel corn with onions and red pepper, sauteed tomatoes and zucchini, and creamed spinach were determined to be a better choice. These vegetables are next reviewed separately and matched with entrees on the basis of suitability.

To match entrees with surrounding items (vegetables and starches, for instance), a menu suitability chart is drawn up as in Figure 5-9. All of the entrees, vegetables, and starches included in the menu repertory are listed on the chart, to provide an easy reference source for the executive chef or manager. Based on the number of entrees to be served in a given meal service, the chef can substitute or combine different items to take advantage of daily specials and seasonally available foods, or compensate for unexpected shortages of certain foods. In addition, the chef can add sauces, such as hollandaise, bearnaise, or veloute, to appropriate entrees to provide extra versatility to the menu. The work page in Figure 5-10 can be used to identify the final menu listing to be used for the printed menu format.

When the first menu items are selected, the chief and restaurant manager take

FINAL MENU LISTING

Experts recommend no more than 14 entree items, 6 appetizers, 3 soups, 4 salads, 4 vegetables, 2 starches, and 6 desserts.

Select items from your completed and evaluated Menu Inventory Forms and fill in the appropriate categories to establish your final menu listing.

APPETIZERS

_____	_____
_____	_____
_____	_____

SOUPS

_____	_____
_____	

SALADS

_____	_____
_____	_____

ENTREES

_____	_____
_____	_____
_____	_____
_____	_____
_____	_____
_____	_____
_____	_____

VEGETABLES

_____	_____
_____	_____

STARCHES

_____	_____

DESSERTS

_____	_____
_____	_____
_____	_____

◆ **FIGURE 5-10.** The final selection of menu items is listed on this work page for evaluation and critique. Reprinted with permission. Copyright 1991 *Menu for Profit.*

one last step to ensure quality control in production over an extended period of time. Specifications for each menu item as it will be presented to customers are drawn up as in Figure 5-11. Each specification includes all of the items that will be included in the presentation. Each item is broken down according to portion size—a 6-ounce filet, one cup of cooked rice, and so on. Garniture is also listed—2 thin lemon slices, 3 sprigs of parsley, or whole curly endive lettuce leaves as a plate liner. The presentation is specified according to the plate size, underliner, or dish. Items are plated following the exact specifications. Photographs are then taken on the item in its ideal presentation state. Figure 5-12 shows a variation of a specification form with photograph and nutritional information. A summary of the specification is then laminated with the photograph and posted at the pickup location in the kitchen and on the line.

Both kitchen staff and servers now have the responsibility of making sure that the items appear on the correct serving dishes with the appropriate garnish and in the specified portion amounts. A consistent complaint of guests who return often to certain restaurants is that a dish they order one night is not the same dish two weeks later. Both profit and customer satisfaction can be seriously affected by inconsistent presentation and portion control. The specifications also act as guidelines

SPECIFICATIONS FOR:		DATE:	
RESTAURANT:			
ITEM NAME:			
INGREDIENTS	PORTION BREAKDOWN	GARNITURE	PRESENTATION

◆ **FIGURE 5-11.** Specification card for individual menu items.

MENU ITEMS: GRILLED TUNA STEAK SANDWICH   USE RECORD #SAN3

SET UP SPECIFICATIONS	INGREDIENTS	PORTION QUANTITY	COST	SET UP DESCRIPTION
14"Glass Oval Duralex Plate	Multi Grain Bread	2 slices		shingle O/Plate frm center-3:00
	Tuna Steak	1 ea/ 5 oz		O/Top bread
	Mixed Greens	3 oz		O/Plate 9:00
	Cucumber, 1/8" slice	3 ea		O/Top mixed green
	Green Pepper Ring 1/8"	1 ea		O/Top cucumbers
	Red Pepper Ring 1/8"	1 ea		shingle O/Top green pepper ring
	Leaf Lettuce	1 ea		O/Plate 2:00 beneath bread
	Tomato, 1/8" slices	2 ea		shingle O/Top lettuce
Hall 844 Ramekin	Mayo Mustard Sauce	1 1/2 oz		in ramekin O/Plate 6:00

♦ **FIGURE 5-12.** This specification form outlines all of the necessary information for the plate presentation of a specific menu item.

to the purchasing agent, who can review them quickly to see what can be affected by price or quality changes. Review the final menu selection using the following questions as one last check for suitability:

1. Are the ingredients available on a year-round basis, at a reasonable price?
2. Will the item pose production problems?
3. Can the product be partially preprepared and finished in a brief period of time?
4. Will the item sell well enough to justify carrying it on your menu?
5. Can you offer the same raw product in a variety of forms?
6. Does the item match the level of cuisine of other menu offerings?
7. Can the item be priced within the range of the other items?

The final menu selection should offer a tempting variety of foods, contrasting in taste, texture, color, and style of preparation. To obtain fresh ideas for menu presentations, the menu planner should consult cookbooks, foodservice industry publications, and other reference sources. In addition, the menu planner should be familiar with all techniques of preparation and production and should know the characteristics of foods from the raw state to the finished product. It is important that the menu planner also understand the holding properties of food items—changes in color, texture, and risk of spoilage when foods are stored at hot or cold temperatures. Research and experience are the best guides to producing creative, successful menus.

CHAPTER 6 ◆

Setting
Quality
Standards

In the foodservice industry, standards are established to produce goods and services of consistent quantity and quality over an extended period of time. The criteria that are used to establish these standards will vary in each operation, depending on the needs of the owner, manager, or corporation. The foodservice industry should set up a specific set of guidelines that can be reinterpreted by individuals to allow for some creativity, but still result in a standard of quality, weight, value or quantity. At present, however, foodservice operations set their own standards in the following eight areas: preparation, production, presentation, purchasing, service, costing, pricing, and quality control.

Standards of preparation and production ensure that when a menu item is prepared, it has the same taste and consistency, and yields the same number of portions every time. Presentation standards ensure that a menu item is always served on the same size serving dish, has the same sauce and/or garnish, and always looks the same. This consistency is called quality control and satisfies the customers, who know that they will receive the same product every time they order it, and pleases management, by guaranteeing the same food cost whenever the food item is produced.

To develop standards of preparation, production, and presentation for a professional foodservice kitchen, three additional standards should first be established:

♦ standard recipes
♦ standard portions
♦ standard yields

A standard recipe is a formulated set of instructions that have been systematically tested and proven to result in a product of consistent quality, when produced under a given set of conditions. A standard yield is the number of portions that will be produced by the standard recipe. This yield is based on a standard portion, the size or measurement of each individual serving. For example, if the measurement of a portion is 4 ounces and the yield is 6 portions, the total liquid measure of the product is 24 ounces or 3 cups. If, however, the portion measurement is sometimes 2 ounces, sometimes 3 or 4 ounces, the yield of the recipe will never be consistent. As you begin costing and pricing, you will see how very important standardization of portions is to the process of costing.

Standard Recipe Cards

To ensure that the standard recipe is followed, and that weights, measures, and directions are easy to read, each kitchen should use standard recipes. Although there are many formats that can be used, the one shown in Figure 6-1 is a good example. There are no absolutes in terms of recipe card formats; they should be adapted to fit your individual needs. Computer software programs offer a variety of formats and reports.

The most important rule to follow is organization. Make sure that the card is properly spaced, that lines separate all sections, and that the print is easy to read. Changing the colors or typeface weight in the headings, for example, will help distinguish the different sections of the card.

CREOLE PORK CHOPS	INDEX #		*Portion:* 1 chop + ¼ cup sauce *Yield:* 50 portions
Ingredients	*Weight*	*Measure*	*Directions*
Pork Chops	16 lb	50 pcs	Trim fat from pork chops Place on sheet pan Brown at 350° for 30 minutes, drain off fat Place chops in baking dishes, 25 chops per pan
Tomatoes Onions, frozen, chopped		1–120-oz can 3 cups	Mix remaining ingredients together and simmer over low heat for 30 minutes
Green Peppers, frozen, chopped		3 cups	Pour 1 quart of sauce per pan over pork chops
Salt		1 Tbsp	Cover pans and bake at
Vinegar		¼ cup	350° (moderate oven)
Worcestershire sauce		¼ cup	1½ hours or until meat is tender

SWEET PEPPER BROCHETTES

Ingredients	Weights	Measures	Directions		Yield: 24 servings
Pesto Marinade	—	—	1. Prepare Pesto Marinade. Reserve.		
Le Rouge Royale® sweet red peppers	4 lbs.	8	2. Rinse peppers. Remove core and seeds. Cut peppers into 1-inch pieces. Cut onions into 1-inch pieces and separate layers.		
Le Jaune Royale® sweet yellow peppers	4 lbs.	8			
Red onions	2 lbs.	4	3. Place chicken, shrimp, and beef in three separate plastic bags. Place peppers, onions, and mushrooms in a fourth bag. Divide and pour marinade evenly into bags. Seal and toss to coat. Marinate several hours or overnight in refrigerator.		
Boneless, skinless chicken breast, cut into 1-inch cubes	2 lbs.	—			
Large peeled shrimp	2 lbs.	—			
London Broil, fat removed, cut into 1-inch cubes	2 lbs.	—	4. For individual servings of 2 skewers each, thread 4 oz. meat, 5 oz. peppers, 1 oz. onions, and 2 mushrooms alternately on 2 bamboo skewers. Barbecue or broil 4-5 minutes per side, turning once.		
Mushrooms, medium size	1 lb., 4 oz.	48			
Bamboo skewers	—	48			

PESTO MARINADE					Yield: Approx. 6½ cups
Garlic	—	8 cloves	1. Place all ingredients in food processor and purée.		
Fresh basil leaves	—	1 qt.			
Olive oil	—	3 cups			
Fresh parsley	—	2 cups			
Lemon juice	—	1⅓ cups			
Salt	—	2 tsp.			
Pepper	—	1 tsp.			

Nutrients per serving based on using Le Rouge Royale® peppers only:

								Percentages of USRDA:	
Calories	432	Carbohydrates	16 g	Fat-Total	30 g	Sodium	313 mg	Vitamin A	38%
Protein	27 g	Dietary Fiber	5 g	Cholesterol	115 mg			Vitamin C	353%

SUN☼WORLD. Foodservice Department, P.O. Box 1028, Coachella, CA 92236   (619) 347-8693

◆ **FIGURE 6-1.**   Standard recipe card formats. The use of standardized recipes ensures that food items will consistently yield the correct number of portions, an important factor in maintaining cost controls.

Index

The index for the menu repertory is numbered by food category, sectioned into courses as they appear on the menu and numbered in the menu repertory. If a sauce, stock, or preprepared ingredient is used more than once in the menu repertory, that ingredient should have its own recipe card. The recipe card for the particular product or dish that uses the preprepared ingredient should indicate the weight or measure of the sauce or stock to be added to the main recipe and refer the cook to the recipe card for the sauce or stock. For example:

Ingredient	Measure	Directions
Sauce Veloute	1 pt	See Sauce Card 5
Chicken Stock	1 qt	See Card 2

To ensure easy accessibility, recipe and cost cards should be cross-referenced. Both sets of cards should have a master index at the beginning of the file that lists exactly which items are available in each category.

Item

Each menu item should be listed on the recipe card in capital letters, phrased as it appears in the menu repertory. If the item is written differently or in another language on the printed menu, the menu listing should follow the main title in parentheses:

SAUTEED VEAL WITH MUSHROOMS
(Saute de Veau aux Champignons)

Yield

The yield is the number of standard portions obtained from the recipe. Each recipe card should be developed so that the number of portions can be easily increased or reduced. A recipe that yields 20 portions, for example, can be multiplied by 2, 4, or 5 to produce 40, 80, or 100 portions. It is also easily reduced by 2, 4, or 5 to produce a recipe for 10, 5, or 4 portions. Do not automatically assume, however, that all ingredients can easily be increased or reduced in the same proportions.

Ingredients

Each recipe card lists the ingredients in their basic state. Limited production directions can be included here, but only as a guide to selection or purchase. For example:

INGREDIENTS
Whole vine-ripened tomato, 6 × 7 flat pack
Spanish onions, 50 lb bag, sliced
Cauliflower, 24 heads

The listing should be as clear and accurate as possible, to ensure that the proper product will be used in the form that you want. If, for a white cake recipe, you listed one ingredient simply as flour, the cook might use anything from all-purpose to whole-wheat flour, when the cake requires white pastry flour. This is also applicable to seasonings and ingredients that can be used in both liquid and solid forms, such as vegetable shortening.

Weight

The only indications that appear in this column are for ingredients that are weighed on a scale. This can be very confusing in the case of ounces.

Ingredient	Weight
Salt	2 oz
Water	

Measure

Any ingredient that is not weighed on a scale appears in this column. This applies to pieces, bunches, cans, and slices as well as standard dry and liquid measures.

Ingredient	Measure
Carrots	1 lb
Celery	2 stalks
Onions, sweet	1/2 lb
Parsley, flat Italian	1/2 bunch
Bay leaf	2 leaves

Directions

Before the recipe card is written, the manager or chef should simplify the directions and place them in the order of production use. Match the directions to the ingredients and section them by lines into an orderly series of directions as in Figure 6-1. Simple directions are best, since space on the cards is limited and complicated directions slow the cook down and increase the possibility of error or misinterpretation, particularly if many different people will be using the cards. Where Spanish is spoken in the kitchen, make sure that the cards are available in both English and Spanish. Good directions increase production organization and facilitate accurate timing and scheduling. There are times, however, when a classical presentation will require advanced terminology and a more complicated series of directions. These are to be applied whenever needed but only when needed.

Standard recipe cards are important. They are developed to provide a consistent standard of portion, yield, and production, thereby assuring quality, quantity, and food cost control. Every item in your menu repertory needs to have a master recipe card. Keeping your recipes in the chef's head will not help your food cost or portion control. In addition, a chef who is here today may be gone tomorrow—with all your recipes.

Establishing Purchasing Standards

To establish purchasing standards for each item, a series of specifications is developed and a food data sheet is compiled. Specifications provide the purchaser with the information needed about the food to be purchased—whether it is fresh, frozen, or canned, for example—and the measure and quality that are desired. The food data sheet, which is a list of every ingredient needed to produce the menu repertory, provides a master purchasing list. There should be a written specification for each item on the food data sheet.

The standard of quality is determined by the established food cost for each item and the overall standard that management sets for the quality of the food that they will serve. Every foodservice operation will determine its own standard of quality and choose its purveyors accordingly.

The standard recipe cards determine the quantities of items to be purchased. The foodservice manager and executive chef use a production sheet to translate the individual amounts used in each standard recipe card into the total amount of each item to be purchased.

Determining the Yield

An important aspect of purchasing is the ability to determine the actual yield from a food product. In addition to indicating the number of standard portions that will be produced by a standard recipe, the term yield also refers to the number of portions that can be derived from a raw food product. Waste, which can be defined as unusable food material, becomes an important consideration in determining the yield of raw food products. Waste takes a variety of forms. In meat, waste consists of bones, fat, trimmings, and evaporation; in fruits and vegetables, waste is trimming and moisture loss. A yield test is conducted to determine the actual edible raw portion, the cost of waste, and the final cost per portion.

In performing a yield test, one food item is tested at a time. The item is tested five times, each time using a food product of the same exact weight, portion or measure, and quality as the other four. Five tests are conducted because at least five repetitions are needed to produce an average result, consistent over a long period of time. If the testing were limited to two or three tests, there would not be enough of a chance for varieties of sizes or properties of ingredients to become consistently apparent. The results of the five tests are totaled and divided by five to give the average overall yield of the product. The multiple-yield test, as outlined in Figure 6-2, can be expensive if conducted on all items in production. If not all items are to be tested, priority should be given to expensive cuts of meat, poultry, and fish. These items would have the highest percentage of waste if produced improperly, without the standard preparation guidelines that are the result of properly conducted yield tests.

To calculate the cost and waste of an individual item, follow the cost yield test format shown in Figure 6-3. The cost per portion and the cost of waste can now be determined by transferring the results from the yield test onto a yield-portion cost format (Figure 6-3). Unit refers to the unit of measure in which the food product was purchased. Waste is the total amount or measure of unusable food. (The waste

MULTIPLE-YIELD TEST FORMAT

TEST #	ITEM		DATE				
COOKING METHOD	COOKING TEMP.		(F)		(C)		
COOKING TIME	PURCHASE WEIGHT						
BREAKDOWN	lb.	oz.	kg.	g.	lb.	oz.	kg.	g.
Bones								
Fat								
Trimming								
Shrinkage								
Storage								
Preparation								
Cooking								
Total Shrinkage								
*Cutting Waste								
Total Waste								
Total Yield								
Portion Size								
Portion Yield								

◆ **FIGURE 6-2.** Suggested format for recording results of multiple-yield tests. Each item is tested a minimum of five times to obtain an average result.

factor was determined by conducting five tests on the food item.) The AP price (as purchased price) is based on the predetermined portion size obtained from the recipe card. Depending on the item being analyzed, this will reflect either the portion to be served or the portion of the ingredient to be used in the recipe. Total extension is total amount of money spent, found by multiplying the AP price by the total number of units per item.

COST YIELD TEST FORMAT

ITEM:_____
WEIGHT:_____lb._____oz._____kg._____g.
PURCHASE COST (total):_____lb._____oz._____kg._____g.
MENU ITEM:_____PORTION SIZE:_____

Breakdown	Weight	Value	Total Value	Cost	*PC	*CF
TOTAL:						

*PC = Portion Cost   CF = Cost Factor

◆ **FIGURE 6-3.** Suggsted format for recording cost and waste of an individual item.

The weight of the edible raw product is converted into ounces because the portion is measured in ounces. The converted figure is now divided by the portion size to find the total number of portions available. The cost per portion is calculated by dividing the total extension by the number of portions:

$60.00 divided by 50 = $1.20

The cost factor per portion provides a quick means of updating individual portion costs when the price of the product changes by pound or volume. The cost per portion (CPP) is divided by the as-purchased price (AP) to determine the cost factor per portion (CFPP). To obtain the final cost per portion, multiply the new AP price by the CFPP.

CPP divided by AP = CFPP

($1.20 divided by $2.00 per lb = 0.60)

CFPP × New AP = Final CPP

(0.60 × $2.50 per lb = $1.50)

Finally, when part of the waste of a product can be reconstituted or sold, this percentage should be given a value and subtracted from the actual waste cost. Fruits and vegetables should not be overlooked when tests are being conducted for food waste. First, the raw product portion must be weighed (here, raw signifies before

APPROXIMATE WASTE IN THE PREPARATION OF FRESH FRUITS AND VEGETABLES			
Fruits	*Average % Waste*	*Vegetables*	*Average % Waste*
Apple	27	Asparagus	49
Avocado	40	Beans, green	16
Banana, peeled	42	Broccoli	38
Blueberries	16	Cabbage, green	21
Cranberries	3	Carrots	24
Cherries, pitted	21	Cauliflower	69
Cantaloupe, with rind	63	Celery	29
Cantaloupe, peeled,		Chicory	29
serving without rind	39	Cucumber	28
Grapefruit, sections		Eggplant	22
and juice	52	Lettuce, head	31
Grapes, seedless	5	Onions, mature	11
Honeydew, served		Peas, green	63
without rind	46	Peppers, green	22
Orange sections	43	Potatoes	16
Peaches (average)	20	Potatoes, sweet	20
Pears	33	Radishes	40
Pineapple	52	Spinach	64
Plums	7	Squash, acorn	34
Strawberries	16	Squash, Hubbard	41
Watermelon	54	Squash, zucchini	2
		Tomatoes	14
		Turnips	20

◆ **FIGURE 6-4.** Charts such as this from the U.S. Department of Agriculture identify waste and help to make costing more accurate.

preparation). After preparation, all material waste is weighed, and the item is cooked and reweighed to determine moisture loss and shrinkage. This loss (moisture and shrinkage) is added to the total weight of the material waste, to provide the total waste factor. Figure 6-4 shows the average percentage of waste of the most commonly used fruits and vegetables. To estimate produce portion costs, if you do not need a precise figure, multiply the weight of the vegetable or fruit by the percentage of waste:

Weight × percentage of waste = actual weight of waste

(6 lb × 0.25 = 1.5 lb)

The method and temperature by which the food is cooked can increase or decrease yield. Meats and poultry cooked at lower temperatures over longer periods of time will suffer less overall shrinkage and have a higher portion yield than meats cooked at a high temperature for a shorter period of time. Meat sizes also affect yield. The larger the piece of meat before cooking, the smaller the shrinkage loss. Braising or searing meats at high temperatures and cooking for the remaining time at lower temperatures will seal the outside tissue and prevent juices from running off in the cooking process.

PRODUCTION SCHEDULE: GERMAN HERITAGE SOCIETY OCTOBER 26

Oysters, Herb Butter	1000	Line: broiler	Banquet box	At service
Mushroom Soup	1000	Range top	Banquet Kitchen	Day before or frozen, reheat prior to service
Green bean salad	1000	Prep kitchen	Walk-in box	Salad: day before Set up: afternoon
Filet of Sole	1000	Banquet oven	Banquet box	At service
Champagne sherbet	1000	Prep kitchen	Pantry freezer	Two days before
Rolled beef	1000	Banquet kitchen	Walk-in box/ banquet oven	Day before, reheat for service
Red cabbage	1000	Canned/ banquet kitchen	Banquet range	At service
German potato salad	1000	Banquet kitchen	Walk-in box/ front line ovens	Two days before, reheat for service
Black bread	1000	Vendor/ pantry area	Pantry area	Day of delivery
Black forest cake	1000	Vendor/ pantry area	Tray racks Portions sliced	Day of delivery Tray up: afternoon

♦ **FIGURE 6-5.** Production sheet for 6 course catering menu.

MENU SELECTION   SPEISE KARTE

AUSTERN KAISER WILHELM
(Broiled Oysters in Herb Sauce)

BERNKASTELLER KURFURSTLAY KAYSER

PILZA SUPPE
(Mushroom Soup)

BOHEN SALAT
(Green Bean Salad)

LIEBFRAUMILCH RHEINSONE KAYES

POSCHIERTES SEEZUNGE
(Poached Filet of Sole)

WEISWEIN UND WEINTRAUBEN SOSE
(White Wine and Grape Sauce)

SORBET WEIS
(Champagne Sherbet)

ROULADEN KRAUTNUSSE
(Rolled Beef with Brussels Sprouts and Nuts)

ROTKOHL
(Red Cabbage)

DEUTSCHES KARTOFFELSALAT
(German Potato Salad)

MICHELOB BIER

SCHWARTBROT
(Black Bread)

SCHWARTZWALDER KIRSCH TORTE
(Black Forest Cake)

KAFFEE
(Coffee)

PFEFFERMINTZ SCHNAPPES
(Peppermint Schnappes)

♦ **FIGURE 6-6.** Catering menu for 6 course German Cuisine menu served to the German Heritage Society in Savannah, Georgia.

Vegetables that are steamed rather than boiled will retain more water, vitamins, and minerals and will shrink less. A 10-pound roast, cooked at 300° F to an internal temperature of 140° F will yield 3.8 servings per pound. If the same roast is cooked to an internal temperature of 176° F at an oven temperature of 300°, it will yield only 3.1 servings per pound. Cooked at 350° to an internal temperature of 140°, the roast will yield 4 servings per pound. Last, with a cooking temperature of 200° and an internal temperature of 176°, the roast will yield 2.6 servings per pound.

◆ PART III ◆

Designing
For
Profit

CHAPTER 7

♦

Marketing
With Menu
Design

In the highly competitive world of restaurant marketing, the physical menu is often the critical identifying factor that relays to the customer the restaurant theme, menu concept, and service style. The menu also prepares the customer for the quality level of food preparation and presentation being offered and can affect the perceived value which the customer has for the overall restaurant experience.

Chain restaurant companies develop graphic menu presentations designed to sell a theme or concept. Menu items are often duplicated between parallel levels of restaurant operations such as casual, family-style operations and bistro concepts. Examples of the ways in which graphics and layout are used to identify these operations from each other are highlighted in the design examples featured in this section.

Independent restaurants often use daily menus as a feature of their operation. Daily menus allow restaurant chefs to offer seasonal and regional foods and dishes that help to set them apart from local competition. Daily menu presentations take advantage of computer software programs and on-property laser printers to create colorful, graphic throwaway menus. Menu cover designs are used to professionally present daily menus and are among the examples included in this section.

From coffee shops to five-star dining rooms, the physical menu plays a feature role in marketing food items to the customer. The customer often selects menu items based on the visual layout of the menu. Marketing specific profitable food items to the customer is achieved by following layout and design principles that direct the reader's eye to predetermined areas on the menu. Menu engineering determines the most profitable and best selling menu items. Menu layout helps to place these items in key design areas to which the customer's eye is quickly drawn. Menu items that are not profitable but must be included to respond to customer demand can be placed in areas of the menu that are less obvious and require more effort to locate. This section will discuss the importance of layout to the success of menu item sales and offer design guidelines.

The two primary elements of menu design are the *visual format* and the written *menu copy*.

VISUAL MENU FORMAT

The visual format, or design, of the menu is made up of the following key components:

♦ physical design format
♦ menu item layout
♦ typeface
♦ illustration and graphic design
♦ paper style
♦ color selection
♦ cover design

How well menu designers and marketing experts have identified and incorporated these components will be reflected in the success of menu item sales. Considerations such as how easily the customer can read the menu, the excitement and interest that colors create, the durability of the paper stock, and the statement the menu cover design makes about the restaurant, are some of the important issues that must be addressed in the development of a successful menu.

Physical Design Format

The overall design of the menu is determined by the style of physical format used. Physical formats for menu design vary in terms of size, shape, number of pages, and panels. Formats are available in a wide range of sizes, from a small 4 × 6 inch card to a 13 × 18 inch size or larger.

The term *panel* refers to a single, unfolded section, such as the outer cover of the menu. A *two-panel* menu has the format of a book with a front and back cover, while a *tri-panel* format resembles a page folded into thirds. These and other physical design formats are shown in Figures 7-1–7-12. Menu designs also vary in shape (rectangular, circular, or triangular, for example) and the number of pages between the panels.

The physical design format of the menu should be selected on the basis of how easily the customer can handle it and how well it will sell the concept and selected menu items of the restaurant. Menus that are too large may knock over glassware or overcrowd the table. Menus with too many pages may confuse the customer with a catalog of choices, limiting their menu selections to items that are familiar to them. Different physical design formats are designed to suit specific menu requirements.

Single Panel

The one-page format shown in Figure 7-1 is used to present limited menus and special selections. Although the most common sizes for single-panel menus are 6 × 8 inches and 9 × 10 inches, designs are also available in larger sizes. The two single-panel menus from Rachel's Kitchen in Figure 7-1 represent the use of this format to present limited menu offerings such as breakfast items. Figures 7-2a and 7-2b from Rhett's use the single panel for a two-sided luncheon menu placement displaying food items on side one and beverage items on side two. Both menus in this section are laminated for cleanliness and increased durability.

Classic Two-Panel Fold

This traditional format is the most popular design for menus offering a classical presentation of courses. Space can be the most significant problem with this format. If too many courses are offered, the layout will be crowded and unbalanced. This format is most appropriate when menu selections are limited and the design of the overall layout is carefully executed. Figure 7-3 is a classical two-panel fold menu featuring the appetizer, salad, and entree courses from Croc'n Berry's Restaurant. In addition, the menu identifies featured combination platters in the lower left corner and regional fish in the upper right. House specials are centered in the middle of the right panel. By including choices of starches and vegetables with all entrees, management has eliminated the need to take up menu space to identify side dishes.

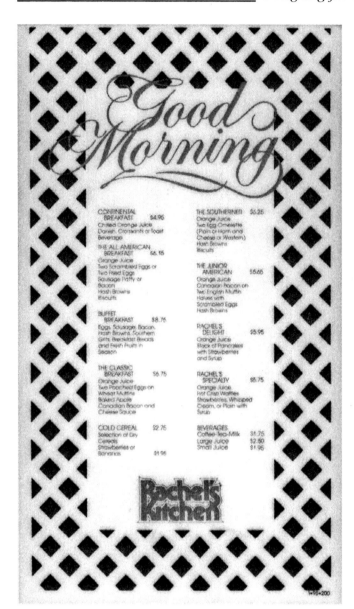

♦ **FIGURE 7-1.** The single-panel menu format is used for both the standard breakfast menu and the limited late morning menu from Rachel's Kitchen in the Opryland Hotel, Nashville, Tennessee. Reprinted with permission of Opryland Hotel, Nashville, Tennessee.

Two-Panel Multi-page

This format provides ample space for menus offering a large selection of items as featured by the Japanese menu in Figure 7-4. As with all physical formats, the number of pages should depend on what the customer can handle comfortably.

Single-Panel Fold

The single-panel fold format shown in Figure 7-5a takes a standard two-panel menu and adds a narrow extra fold to the right side of the menu. When the menu

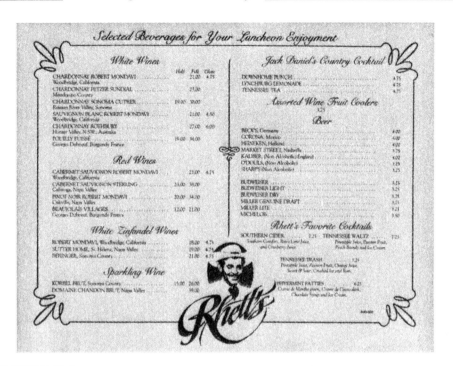

♦ **FIGURE 7-2a.** The placemat menu in these two figures illustrates an optional application of the single-panel format. Rhett's Restaurant, Opryland Hotel, Nashville, Tennessee. Reprinted with permission of Opryland Hotel, Nashville, Tennessee.

♦ **FIGURE 7-2b.** The reverse side of Rhett's placemat listing alcoholic beverage selections.

♦ **FIGURE 7-3.** The classical two-panel fold format identifies the menu selections for Croc'n Berry's. Reprinted with permission of Ad Art Litho.

♦ **FIGURE 7-4.** A two-panel multi-page format is used for this Japanese menu from the Hotel Nikko, Manila. Reprinted with permission of Dusit Hotel Nikko, Manila, Philippines.

♦ FIGURE 7-5a. A narrow, side-panel fold is added to this two-panel format to extend the available space for menu items without creating a full third panel. Reprinted with permission of Ad Art Litho.

is fully opened, the side panel includes a location for a daily specials insert along with additional menu items. Figure 7-5b shows the side panel folded inward to reveal the dessert and beverage menu.

Horizontal Two-Panel Fold

This format type is simply an adaptation of the classical two-panel fold menu. The horizontal shape offers an alternative way of opening and displaying menu items, creating the effect of a single menu, as seen in the menu from Harley Davidson Cafe in Figure 7-6. If this format does not become too large, it is easy for the customer to handle.

This design can also be expanded to a multi-page format. The major problem with adding pages to this format is that they must be securely fastened or pages will slip out and become disorganized, creating a sloppy and confusing presentation to the customer.

Appetizers

HOMEMADE SOUP OF THE DAY	1.45	FRIED BREADED MUSHROOMS	1.75
ONION RINGS	1.75	MOZZARELLA STICKS	3.25

TOMATO, ORANGE or GRAPEFRUIT JUICE   Regular .80   Large 1.05

Kim's Sandwiches & Burgers

1/4 LB. CHEESEBURGER	2.65
1/4 LB. HAMBURGER	2.45
KIM'S TURKEY CLUB SANDWICH	4.25
BACON, LETTUCE & TOMATO	3.25
GRILLED CHEESE SANDWICH	2.45
CHICKEN PATTY SANDWICH	3.75
FISH SANDWICH	2.75
BROILED CHICKEN BREAST	3.75
DOUBLE CHEESEBURGER	3.65
PATTY MELT with Grilled Onions	3.25
PERCH SANDWICH	4.45
HAWAIIAN CHICKEN SANDWICH	3.95

KIM'S TURKEY CLUB DELUXE 5.45
With French Fries and Cole Slaw.

1/4 LB. CHEESEBURGER DELUXE		1/4 LB. HAMBURGER DELUXE	
With French Fries & Cole Slaw	4.45	With French Fries & Cole Slaw	4.25

ADD-ONS
BACON 1.00   MUSHROOMS .45   EXTRA CHEESE .25

Soup n' Sandwich Combos

GYROS	3.95	TUNA SALAD SANDWICH	3.75
GRILLED HAM & SWISS	3.75	CORNED BEEF ON RYE	4.75
CHICKEN SALAD SANDWICH	3.75	REUBEN SANDWICH	4.75

Above Sandwich Combos Served with Cup of Soup and Potato Chips.

Hot Open-Faced Sandwiches

HOT ROAST BEEF	4.75	HOT ROAST TURKEY	4.75
HOT MEATLOAF	4.55	HOT VEAL CUTLET	4.75

Served with Mashed Potatoes and Gravy.

Salads & Cold Plates

CHICKEN CAESAR SALAD	6.45	PEACH or PINEAPPLE PLATE	
CAESAR SALAD	4.65	With Cottage Cheese and Raisin Toast	4.45
GREEK SALAD	4.95	COLD CHICKEN or TUNA	
CHEF'S SALAD	4.95	SALAD PLATE with Potato Salad	4.75
KIM'S GRILLED CHICKEN SALAD	5.95	TUNA or CHICKEN SALAD	
		With Tomato Slices	4.45

Served with Dinner Roll or Crackers.

SALAD DRESSINGS: House (Creamy Italian with Feta Cheese), Creamy Italian, Golden Italian, Fat-Free Italian, French, Thousand Island, Sweet 'n Sour and Ranch. Bleu Cheese Dressing .25 extra

Gift Certificates are Available.

Kim's Specia[l]

CHICKEN KABOB over Rice	5.95
ROAST BEEF DINNER	
With Whipped Potatoes and Gravy	5.95
CHICKEN CROQUETTES with	
Whipped Potatoes & Cream Sauce	5.75
BABY BEEF LIVER & ONION	
With Mashed Potatoes and Gravy	5.45
CHICKEN FRIED RICE	5.95
GOLDEN FRIED CHICKEN	
With French Fries	5.95
HOMESTYLE MEATLOAF	
With Whipped Potatoes and Gravy	5.45
BREADED VEAL CUTLET	
With Mashed Potatoes and Gravy	5.25
TURKEY DINNER with Mashed	
Potatoes, Gravy and Cranberry Sauce	5.95
KIM'S VEAL DINNER	
With Whipped Potatoes & Gravy	5.65

Kim's Specialty Dinners Served with
Applesauce or Cole Slaw. Di[...]

Classic Itali[an]

VEAL PARMIGIANA with Spaghetti	6.25	S[...]	W
CHICKEN ALFREDO	5.95	S[...]	W
FETTUCCINE ALFREDO	4.95	W[...]	

Pasta Specials served with Cup of So[...]

From the [...]

CHOICE PORTERHOUSE STEAK	11.95	C[...]
RIB EYE STEAK	8.95	C[...]
B.B.Q. RIBS	6.95	C[...]
		W[...]
B.B.Q CHICKEN BREAST	5.95	B[...]
HAM STEAK with Pineapple	5.75	& [...]

Served with Cup of Soup, Choice of Si[...]
and Potato of Your Choice. Di[...]

Seafo[od]

FRESH LAKE ERIE PERCH	7.95	W[...]
FISHERMAN'S PLATTER	7.65	G[...]
FRIED SCALLOPS	6.25	F[...]

Seafood Dinners Served with C[...]
Choice of Potato and [...]

Homemade Desserts

CHEESECAKE 1.75

BAKLAVA		1.55
RICE PUDDING		1.55
ASSORTED FRUIT PIES	per slice	1.65
PIE A LA MODE		2.15
CREAM PIES	per slice	1.65
CAKE		1.55
CAKE A LA MODE		2.05

*From the
Ice Cream Fountain*

SUNDAES		
Chocolate, Strawberry,		
Hot Fudge or Butterscotch		1.95
MILKSHAKES		
Strawberry, Vanilla or Chocolate		1.75
DISH OF ICE CREAM		
Choice of French Vanilla,		
Chocolate or Strawberry		1.25

Beverages

COFFEE or SANKA		.90
HOT TEA		
& DECAFFINATED TEA		.90
HOT CHOCOLATE		.90
SOFT DRINKS	.95	1.05
LEMONADE	.95	1.05
ICED TEA with refills		1.05
MILK	.80	1.05
CHOCOLATE MILK	.90	1.15
MILKSHAKES		
Vanilla, Strawberry		
or Chocolate		1.75

◆ **FIGURE 7-5b.**   When the side panel is folded inward, the dessert listing appears. This maximizes the use of menu space, eliminating the need for a separate dessert menu.

Multi-Panel Fold

Figure 7-7a is a tri-panel fold format. The wide center panel features main course items and sandwiches, while the two side panels identify appetizer, salad, dessert, and beverage items. Figure 7-7b shows the back of the center menu panel and the two side panels closed to complete Roadster's menu cover.

Figure 7-8 limits the menu item selection to three equally sized menu panels. As in Figure 7-3, by including side dishes with each entree, the need to take up additional menu space for a limited salad selection, vegetables, and starches is eliminated.

♦ **FIGURE 7-6.** The top fold on this menu design creates the effect of a single panel when opened. Reprinted with permission of Harley Davidson Cafe, New York, New York.

♦ **FIGURE 7-7a.** A tri-panel format is used to display the wide variety of menu items at Roadsters in Lewes, Delaware. Reprinted with permission of Roadsters Restaurant, Lewes, Delaware.

♦ **FIGURE 7-7b.** The back of Roadsters' menu includes two side panels that close to make the cover and a center back panel, used for additional menu information.

◆ **FIGURE 7-8.** A three-panel format from Longhorn Restaurants. Reprinted with permission of Longhorn Restaurants.

The menu for Sam's Cafe in Figure 7-9a is presented as a single cover panel which opens, as shown in Figure 7-9b, to offer the appetizer and salad courses on the inside left panel and a translation of Spanish menu terms on the reverse side of the third panel. In Figure 7-9c the entire menu is revealed, as the third panel is turned back to list the entree selections, with pasta, tacos, and sandwiches featured in the center panel. As in Figure 7-8, the menu selection is limited to the three equally sized panels on one side of the menu only.

A horizontal variation of the tri-panel menu is shown in Figure 7-10a. Chili's menu opens across the center to simulate raising the cover of a barbecue grill. Menu items, as seen in Figure 7-10b, are laid out horizontally in three separate panels. As in Figure 7-6, the final design effect is a single-panel menu. The difference between the design and layout of these two menus will be discussed later in this section.

The multi-panel format in Figures 7-11a and 7-11b uses a total of six panels and the tri-panel design to display a complete menu selection from breakfast to dessert. The back of the center panel is used for the breakfast menu, which can be offered to the customer as a single panel. The cover and dessert menu take up the other

♦ **FIGURE 7-9a.** This single-panel menu cover is part of a multi-panel format. Sam's Cafe, Scottsdale, Arizona. Reprinted with permission of Sam's Cafe, Scottsdale, Arizona.

two back panels. Lunch and dinner items combine on the front side of the three panels. Entrees and hot sandwiches are featured in the center panel.

A multi-page menu with a spiral binder uses staggered page sizes to create a menu index in Figure 7-12. This format is a good alternative to an oversized multi-panel menu for restaurants that offer an all-day menu with a large number of selections.

Design Variations

In addition to the standard formats, designs can be as varied and creative as menu offerings themselves. Many of these designs require a process known as die cut, which is considerably more expensive than standard panel menu formats. Figure 7-13 shows an unusual two-panel multi-page format. The die cut design of a chef represents the Italian national colors, red, green, and white, reflecting the Italian cuisine theme of the restaurant.

◆ FIGURE 7-9b. These two panels are revealed as the menu for Sam's Cafe is opened.

◆ FIGURE 7-9c. The complete menu is revealed as the Spanish glossary panel is turned back.

♦ **FIGURE 7-10a.** The cover of a tri-panel menu with a top and bottom fold.

Figure 7-14 is a cut-out design which folds into the shape of an ice cream cone and features a limited sandwich and ice cream menu along with beverages for an ice cream fountain snack shop. The menu in Figure 7-15 is the outline of the lower leg of a basketball player that is the signature of the Wilt Chamberlain Restaurant.

Hard Rock Cafe's menu in Figure 7-16a uses the outline of a guitar for the right side of the front cover. The menu layout, seen in Figure 7-16b, carries through with the guitar outline design. The back cover protects the guitar outline of the front cover from being damaged. The entire menu is laminated to extend the life of this expensive menu design.

The promotional menu for Applebee's, seen in Figure 7-17, is completely cut out in the shape of a western boot. Menus such as this have limited shelf life. Continual use will break down the toe and heel of the boot, even if heavier paper stock and lamination are used. The sunglasses design in Figure 7-18a was created with the expectation of limited use for a pool-side snack bar. The rounded edges of the design, however, will stand up longer than the angular corners of Figure 7-17. The menu opens, as seen in Figure 7-18b, to show that the design is carried through on both the front and back panels. The menu layout takes advantage of the outline, offering food items on the lower half of the panel and beverages on the top half.

The menu cover design in Figure 7-19 includes a clear plastic panel painted with a stained glass window design. As the menu cover opens, light comes through the panel, accentuating the design. The window panel reflects the interior design of the restaurant by duplicating the stained glass window panels on the exterior wall of the dining room. While this is an expensive investment, the overall design is conven-

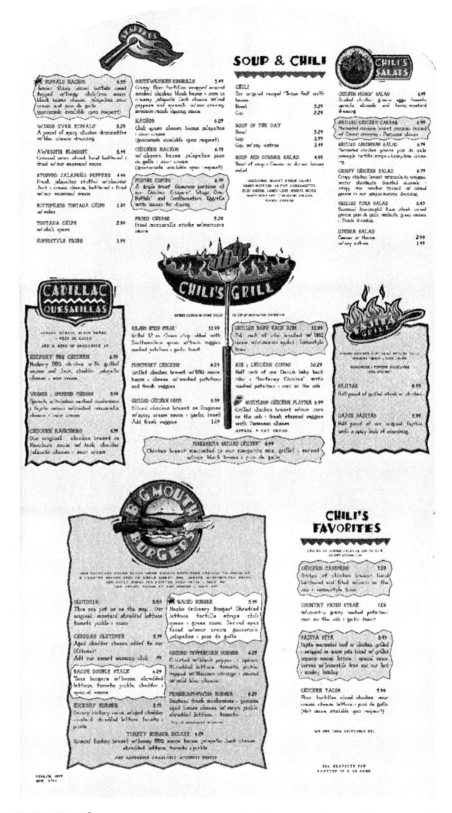

♦ **FIGURE 7-10b.** The menu opens to display a single panel layout. Chili's Grill & Bar restaurants.

125

◆ **FIGURE 7-11a.** A total of 6 panels make up this multi-panel format. The interior menu display identifies the all-day menu. Reprinted with permission of Ad Art Litho.

◆ **FIGURE 7-11b.** Two panels on the back of the menu list dessert and breakfast items. The complete menu folds with the cover as the front panel. Blake's restaurant. Reprinted with permission of Ad Art Litho.

♦ **FIGURE 7-12.** A multi-page menu with a spiral binder using an index created by staggered page sizes. Reprinted with permission of Ad Art Litho.

♦ **FIGURE 7-13.** A two-panel multi-page menu design with pages of different widths so that the die cut shape of a chef serving pizza is repeated when the menu is closed. Reprinted with permission of Il Pastaio, Philadelphia, Pennsylvania.

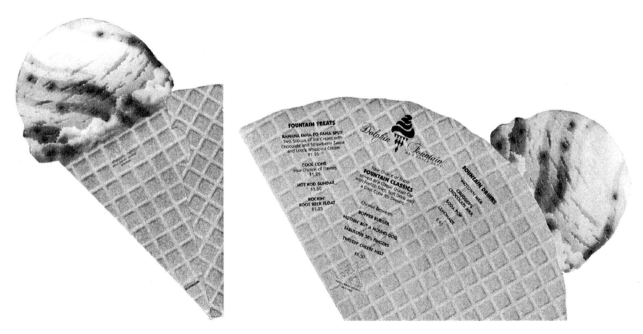

♦ FIGURE 7-14. A multi-panel cut-out design that folds into an ice cream cone shape. The Swan and Dolphin Hotel, Orlando, Florida. Reprinted with permission of Walt Disney World.

tional and will stand up well to repeated use. In addition, a heavy, coated paper stock is used to increase durability.

Figure 7-20a is a complicated and dramatic die cut design showing the center panel of the menu as it is presented to the guest. Figure 7-20b displays the menu panels as they open. The left-hand panel lifts, showing the cut-out balcony design and revealing the next panel, which features a narrative history of the building and surrounding gardens. The complete menu is a tri-fold presentation with inserts for daily menus on the two remaining panels. Unfortunately, this menu will not hold up well under continued use. The cut out of the balcony, while protected, will be handled by the customer and get caught on menu corners in storage.

A menu design is created to suit a specific need. Unless the design fulfills the need, the design is flawed. Menu formats should be constructed around two primary concerns:

1. Can the customer handle the menu comfortably? Is it too large, too small, or too complicated?
2. Will the menu stand up under continued use? Is the design functional, or will it fall apart? Is the paper strong enough to support the design? Will pages fall out or become disorganized?

Menu design and printing costs represent a substantial investment for any foodservice operator. Skimping on the design process or using cheap materials will cause the menu to be redesigned and/or reprinted, creating unnecessary and additional expense for an operation.

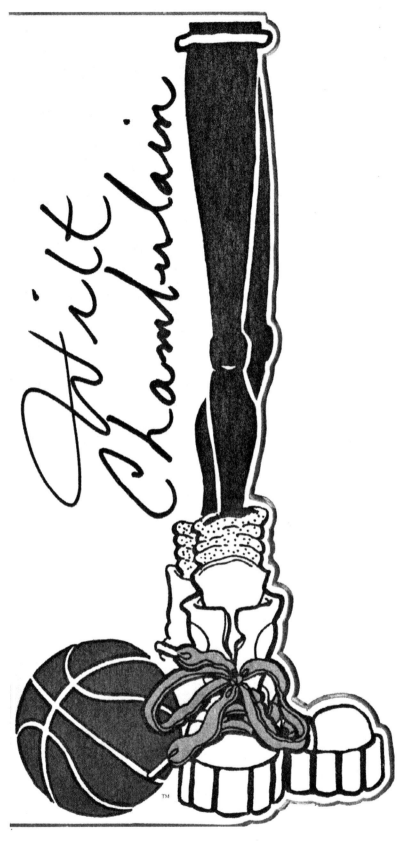

◆ **FIGURE 7-15.** An oversized menu design uses the outline of the lower leg of basketball player Wilt Chamberlain as the menu cover. Reprinted with permission of Wilt Chamberlain's Restaurant, Boca Raton, Florida.

◆ **FIGURE 7-16a.** A die cut outline of a guitar is the cover for Hard Rock Cafe's menu. Reprinted with permission. Hard Rock Cafe is a registered trademark of Hard Rock Cafe Licensing Corporation.

◆ **FIGURE 7-16b.** When opened, the menu is presented in a two-panel layout.

◆ **FIGURE 7-17.** A side panel fold anchors the die-cut outline of a western boot for this promotional menu from Applebee's restaurants. Reprinted with permission of Applebee's Restaurants.

♦ **FIGURE 7-18a.** Sunglasses are the cut-out design for this two-panel, verticle fold menu. Cabana Bar & Grill, Orlando, Florida. Reprinted with permission of Walt Disney World.

♦ **FIGURE 7-18b.** The sunglasses open to display a single-panel menu format. Reprinted with permission of Walt Disney World.

♦ **FIGURE 7-19.** An elegant menu cover design includes a clear plastic panel with a stained glass window design. Reprinted with permission of Hotel Hershey, Hershey, Pennsylvania.

♦ **FIGURE 7-20a.** The Hotel Hershey luncheon menu is a three-panel fold that combines a die cut design and photographs for the menu cover. Reprinted with permission of Hotel Hershey, Hershey, Pennsylvania.

Luncheon Menu

Welcome to a dream come true! Milton and Catherine Hershey traveled the world together, experiencing some of the finest spas, hotels and resorts found in Europe and the Middle East. These were from the golden age of elegance and excellence . . . from the years of the "Grand Tours" . . . from the waning age of aristocracy when crowned heads of Europe were likely to be found at the next table. Together, they dreamed of creating such a luxury hotel in the United States . . . in Hershey. Welcome to their dream . . . enter from the grand Mediterranean Courtyard and relax with a 4-star luncheon in the legendary Circular Dining Room. Take a moment to gaze upon the beautiful and restful formal gardens and reflecting pools just outside the 13 original handcrafted stained and painted windows. Enjoy your visit with us, and share for a time the dreams of Milton and Catherine Hershey.

♦ **FIGURE 7-20b.** When opened, the design reveals an open balcony looking over the terrace pool. Reprinted with permission of Hotel Hershey, Hershey, Pennsylvania.

Menu Item Layout

Layout is the placement of menu items on the surface of the menu format. In general, items are presented in the order in which customers will select them. Formal restaurants will adhere strictly to course sequence with their menu layouts while casual, family-style operations use design, illustration, and color to stimulate the customer and organize the menu on a more casual theme. Most foodservice operations will, however, follow the standard outline of courses outlined in Table 7-1, and discussed in Chapter 1, according to the cuisine and culture being featured.

The layout of the final course structure is an important part of marketing the menu. How easily the customer can select items depends a great deal on how much thought is given to the layout and the design of the area in which special items are featured. Figure 7-21 shows the menu from a casual, family-style restaurant covering lunch through late night snacks. Although the layout of this tri-panel menu is crowded, every effort has been made to create a menu that will generate maximum sales. The center panel design draws the customer's eye and features entree items and specialty sandwiches. Appetizers, soups, and salads are located on the inside cover panel, while burgers and desserts—along with an area for daily specials—are placed on the third panel. Section headings make it easy to identify each of the menu categories in the overall menu design and create a sense of organization in what could be a confusing list of menu items.

The two-panel format in Figure 7-22 presents appetizers and entrees. The outlined box in the center of the second panel identifies signature fish dishes. The menu layout includes a small section at the bottom of panel two offering side dishes and an explanation of nutritional notations for each menu item at the bottom of panel one. As in Figure 7-21, the section headings clearly define the different areas of the menu for the customer.

The menu for Cracker Barrel restaurants shown in Figure 7-23 includes a large number of menu items. This is a tri-panel menu that uses graphically outlined boxes to identify sections of the menu and keep the menu from becoming jumbled and confusing. Layouts should be designed to highlight high-profit items. The size of the top section on the center panel coupled with the graphic outline design draws the customer's attention and features the highest priced main course items. Panel one lists salads, soups, and sandwiches, with a very small section at the bottom of the panel to support the all-day availability of breakfast items. The highest priced

◆ TABLE 7-1 ◆

Standard Outline of Menu Courses

Lunch	Dinner
Appetizer	Appetizer
Main Course (Entree)	Salad
Dessert	Entree
	Dessert

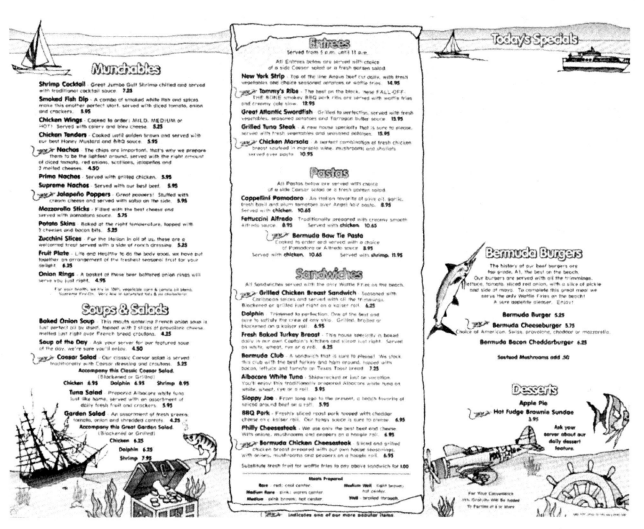

♦ **FIGURE 7-21.** A tri-panel menu layout includes a specials box in the upper right corner. The Bermuda Triangle, Ft. Lauderdale, Florida. Reprinted with permission of Ad Art Litho.

menu items are in the middle of the third panel. As sales for these items will not be significant in the overall sales mix, placing them in this area of the menu makes them visible and available, but does not use up prime marketing space on the overall layout. Side items and lower priced entrees along with beverages get the least visible space, the bottom section of the right hand panel. The customer volume in Cracker Barrel restaurants requires that customers order as quickly as possible to increase table turnover rates.

Layouts should be designed so that the menu is clear and easy to follow. Many menus have attached tip-ons or extra menus clipped over the printed menu. This kind of presentation can be confusing and messy, inhibiting sales. Specials that are offered on a daily or frequent basis should have a clearly identified location in the menu layout, such as in Figure 7-21.

For menu designs such as shown in Figure 7-24, basic layout principles must still

◆ **FIGURE 7-22.** A two-panel layout featuring section headings to clearly identify menu areas. Harry's Safari Restaurant, Orlando, Florida. Reprinted with permission of Walt Disney World.

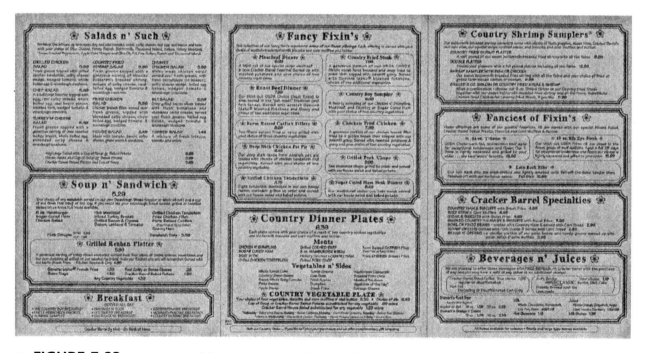

◆ **FIGURE 7-23.** A tri-panel layout uses graphics to clearly outline menu sections for the customer. Reprinted with permission of Cracker Barrel Restaurants.

♦ **FIGURE 7-24.** A verticle, two-panel fold is the design for a seasonal menu for Hard Rock Cafe. Reprinted with permission. Hard Rock Cafe is a registered trademark of Hard Rock Cafe Licensing Corporation.

apply. This two-panel, horizontal fold design is a seasonal menu for Hard Rock Cafe. The top panel identifies the featured appetizer and desserts of the season, while the bottom panel highlights main course items. According to design principles, the customer's eye will be drawn first to the item at the top of the layout—turkey feast on a bun, then to the far right—pot roast, next to the far left—meatloaf, and finally, to the bottom section—chicken salad. In order to encourage or discourage sales of particular menu items, the layout should place items in sections of the menu according to how quickly the customer will find them in the overall layout. Items that have to be searched for can be completely overlooked, such as the desserts in Figure 7-21 and the signature sides in Figure 7-22, and the breakfast section in Figure 7-23. Table 7-2 presents the order in which it has been determined that a reader's eye will travel to specific areas of a page layout for all types of visual materials. The placement of menu categories and/ or specific items in these areas is part of the overall marketing effort for a foodservice operation. As noted earlier, the sales of menu items can be increased or decreased according to how they are placed on the overall menu layout.

The menu examples in Figure 7-25 show the application of these theories. Often, as noted in these examples, typeface and graphic design help to manipulate the reader's eye to specific locations and menu items.

In the single-panel breakfast menu, the eye immediately focuses on the outlined box for fast breakfast specials in the center of the top panel. Along with location, color, typeface, and graphics help to direct the customer's attention to this section. The eye travels upward to the outlined buffet promotion, upward to read the two items directly under the breakfast heading, and then turns down, traveling to the lower half of the menu, where it will first settle on the section's griddleties and eggs & things. The items in the bottom section of this panel will be the last to catch the customer's attention visually.

The two-panel menu uses typeface size and thickness along with graphics to draw the reader's eye. The center of the right side panel is the first visual stop on this menu, with the featured Cuisine of the Americas items. Traveling upward through the sandwich listing, the eye then turns and travels over to the top of the opposite panel and stops at the appetizer and soup section. It then follows the menu panel down through salads and pastas before moving back to the entree section at the bottom of the right panel. Having looked over this section, the eye will usually continue back over to the middle of the opposite panel, stopping on the salad section in this example.

As the multi-panel menu format requires opening the two side panels, the viewer initially treats the center panel as a single-panel menu, focusing on the center section of the middle panel. Then, as in the two-panel format, the eye travels to the section of the middle panel. The eye then, again as in the two-panel format, travels to the top of the third panel, where profitable daily specials are listed on a card with an adhesive backing. The viewer then reads across the top of the center section to the appetizer listing, down the first panel to soups and salads, across through the sandwich section in the center panel to focus on the copy for the burger selection. The lower part of the third section is the last to get the customer's attention. In this case, desserts occupy this space, which is an appropriate use of marketing space for this menu program. As with the single-panel menu, color, illustrations, and typeface play a major role in directing the viewer to specific areas of the menu.

◆ TABLE 7-2 ◆

Visual Menu Item Identification

Single-Panel Menu

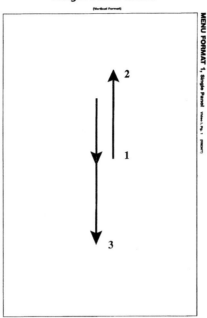

Two-Panel Menu

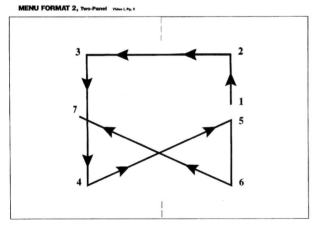

Tri-Panel Menu

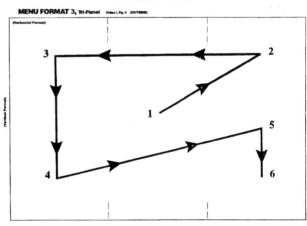

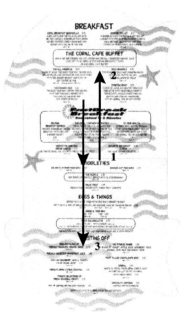

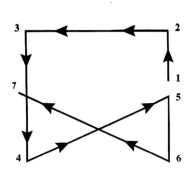

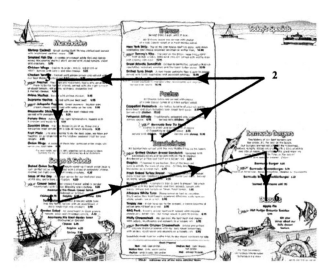

◆ **FIGURE 7-25.** The reader's eye will travel in different directions depending on the overall design and layout of the menu, single-panel, two-panel, or multi panel.

Goudy Cursive
abcdefgghijkklmnopqrstuvvwvwxxyz
AABCDEFGHIJKKLMNNN
OPQRRSTUVWXYZ
1234567890 (&.,:;!?"-$) Thffffl

Bodoni
abcdefghijklmnopqrstuvwxyz
ABCDEFGHIJKLMNOPQRSTUVW
1234567890 XYZ
(&.,:;!?""--$¢%£) ffflffiffl

Bodoni Open
abcdefghijklmnopqrstuvwxyz
ABCDEFGHIJKLMNOPQRSTUVW
1234567890 (&.,:;!?"-$¢%£) XYZ

♦ **FIGURE 7-26.** Contrasting typefaces.

Typeface

The typeface or style of lettering selected for the menu will have a significant effect on the customer. Contrasting typefaces are used to highlight high-profit items or sections to promote sales as in Figure 7-26. How well the customers can read the menu will influence their menu selections. In addition, different typefaces convey different moods. The style of typeface that the menu planner or designer chooses should express the character and personality of the restaurant and be consistent with the overall menu design.

Typography is the general term for the mechanical art of printing; typeface refers to the style of lettering or type used in the printing process. Each letter, symbol, or number is represented by a piece of type. The word appetizer, for instance, requires that nine separate characters of type be set. Hundreds of different typefaces are available in a wide array of styles. The typefaces shown in Figure 7-26 illustrate just a few of the commonly available styles. Computer software programs focusing on graphic design offer a wide selection of typefaces, in addition to advertising and design reference books available in local libraries and bookstores.

Several characteristics distinguish typefaces from each other. Type weight refers to the degree of thickness and boldness of the typeface. The typeface Helvetica Thin shown in Figure 7-27, for example, is lighter and more slender than the bolder Zapf Chancery (Figure 7-27). Roman type is the usual upright form of a typeface, while italic type is the slanted version. Most typeface styles are offered in a number of

♦ **FIGURE 7-27.** Six commonly used typefaces: Commercial Script, Helvetica Thin, Zapf Chancery Medium Italic, Goudy Cursive, Bodoni, and Bodoni Open.

different sizes, weights, and variations on the original style, such as these examples of univers typeface; _Italic,_ outlined, **shadow**. _CG Times, a common word processing font, is another example of how to create **variety**_ within a particular style. Compare the Bodoni Open with the standard Bodoni in Figure 7-27. Visualizing typefaces for use in menu copy requires design knowledge and imagination. The samples of typeface in Figure 7-27 offer a variety of type styles used for headings, menu items, and descriptive copy. Type is used on three different sections of the menu—the headings, the menu items, and the copy.

Headings
are the titles of the courses or names of food categories that divide the menu into sections, such as appetizer, side dishes, seafood, beverages.

Menu items,
of course, are the individual dishes within each course or category, such as prime rib of beef or rice jambalaya.

Copy
is used for either descriptive or marketing purposes. Descriptive copy is the written description of each menu item, while merchandising copy sells other features of the restaurant, such as historical information, special events, or catering and private party facilities.

Deciding What to Sell

Before the menu planner or designer determines how to use typefaces effectively, management identifies the sections and/or menu items they want to have the highest volume sales. Typeface can be used to direct the customer's eye by highlighting specific menu areas. The use of a distinctive boldface type for the headings emphasizes the sectioning of the menu without the use of illustration or graphic design as in Figure 7-28. Figure 7-29 on the other hand, presents the same typeface style in three different weights as seen in the menu from Pusser's Restaurant.

Choosing the Typeface

The choice of lettering styles for each section of the menu is more than a preference for particular designs. For our purposes, the type specifications that must be considered in selecting a typeface can be divided into ways in which legibility is affected and techniques used to highlight or feature special sections of the menu. The three considerations that affect legibility are spacing, contrast, and uniformity of design.

Spacing   Each piece of type has its own dimensions of size and weight. *Letter spacing,* the amount of room between each letter of the word, can affect the legibility of the type, as can *word spacing,* the amount of space between each word. While the designer can request that a typeface be set with loose, regular, or tight letter spacing or word spacing, space is often at a premium in menus and words must be set close together. One major rule should be adhered to: letters that are spaced closely together in long sections of copy should be simpler in typeface style than those spaced widely apart. Italicized type, in particular, can be difficult to read in lengthy sections, depending on the typeface. If italicized type is chosen for long parts of copy, the style should be plain rather than excessively ornamental.

Contrast   A crowded menu can appear cluttered and be difficult to follow, resulting in lower sales than would be realized from one that is well laid out. Contrast is a technique in highlighting. The use of a variety of colors and typefaces along with blank, or white, space can make the menu much easier to read. Separate menu sections should be easy to distinguish from one another, even if the entire menu is to be equally emphasized.

One way of providing contrast is to surround a letter or word with a significant amount of white space. This is called *lightening.* The more lightening used around a letter, the more it will stand out on the menu. Using a contrasting color of type against the white space can heighten this effect. The designer should be careful not to overdo lightening, as letters can appear disconnected, defeating the original purpose of legibility.

Uniformity of Design   While contrasting typefaces, colors, and lightening can make the menu easier to read, all styles and colors should be compatible with one another. Using incompatible, unmatched components will result in a disorderly, confused menu layout. The overall design should express uniformity. While style choices are certainly a matter of individual taste, selecting widely divergent typefaces—for example, using an ultramodern typeface with a formal, traditional one—will most likely produce a startling but negative effect on the menu reader.

Appetizers

LOBSTER MEDALLIONS "NOUVELLE"
Served on Fresh Spinach, Walnut Oil Dressing

SMOKED SALMON
Garnished with Horseradish, Capers, Onions, Olives

MARYLAND LUMP CRABMEAT
Accompanied by Grapefruit and Orange Segments

ASSORTED HORS D'OEUVRES
Smoked Salmon, Prosciutto Ham, Crabmeat, French Pate

JUMBO GULF SHRIMP COCKTAIL

FRESH ARTICHOKE BOTTOM "PRINTANIER"
Sauce Hollandaise

IMPORTED SNAILS "AMERICAN"
Sauteed with Garlic and Herbs, Served in a Squash

BAY SCALLOPS "SURPRISE"
Safran Cream Sauce

FETTUCCINE "ALFREDO" (for two)
Prepared Tableside, Per Person

Soups

LOBSTER CREAM "CARDINAL"

ESSENCE OF ESCARGOTS "ALSACIENNE"

FRENCH ONION SOUP AU GRATIN

CHILLED TOMATO BISQUE WITH GIN
Prepared Tableside

Salads

METROPOLITAN CLUB GOURMET SALAD
Crisp Hearts of Romaine and Boston Lettuce,
Toasted Sesame Seeds, Romano Cheese, Fine Herb Dressing

SPINACH SALAD
Tossed with Fresh Mushrooms, Eggs, Bacon and House Dressing

SALAD "MELE"
Sliced Tomato, Avocado, Crisp Lettuce and Cucumbers

CAESAR SALAD (for two)
Crisp Romaine, Anchovies, Parmesan Cheese, Blended to Perfection
Specially Prepared Tableside, Per Person

Vegetables

ASPARAGUS SPEARS, SAUCE HOLLANDAISE

SAUTEED FRESH MUSHROOMS

FRENCH FRIED ZUCCHINI, SAUCE BEARNAISE

GREEN BEANS WITH BACON

BRAISED CELERY HEARTS

FRESH CAULIFLOWER "POLONAISE"

For the Light Appetite

Served with the Metropolitan Club Gourmet Salad
and Garnished with the Vegetable du Jour

ROAST PRIME RIB EYE OF BEEF

BROILED FILETS OF SOLE

PETITE FILET MIGNON, SAUCE BEARNAISE

BONELESS BREAST OF CAPON

♦ **FIGURE 7-28.** This menu from the Metropolitan Club in Houston, Texas, used a bold typeface set off from the rest of the menu for the headings, effectively identifies each section without the use of illustrations or lines. Reprinted with permission of Ad Art Litho.

◆ FIGURE 7-29. This menu uses the same typeface in three different weights. Reprinted with permission Pusser's at the Beach, Ft. Lauderdale, Fl.

In addition to producing a clear, legible menu, management will frequently decide to highlight or feature special sections of the menu to promote sales. Various techniques can be used to emphasize or lend greater importance to specific areas.

Emphasis   Large menus are often at an advantage for the practice of highlighting since they generally have more available space. When space is not at a premium, the designer can use large display type for heading, titles, or initial letters. _Display type_ refers to the largest sizes of type and differs from _text type,_ which is smaller and used for copy. Although display type is not limited to any particular style, decorative, script, and cursive styles are most frequently chosen, because they stand out more than plainer typefaces.

In general, large type is more significant than smaller type, while boldface type dominates a thinner, lighter typeface. Italicized type gives emphasis, particularly when used in the middle of a sentence or section, but is not as prominent as the use of uppercase letters.

Because the menu relies so much on the printed word to sell food and beverage products, the typeface selected deserves professional consideration. A menu can fail if customers cannot read menu items easily and if the menu is not visually stimulating.

Paper Quality

Paper is an important production consideration in the design of the physical menu. The type of paper chosen will affect the following menu characteristics:

Overall size and format design of the menu
Use of color and illustration
Durability
Printing process expense

Paper has a number of characteristics that can make it more or less suitable for commercial use. Knowing some basic facts about paper will help the menu designer determine which paper is best for the design and will make it easier to work with printers to achieve the best product at the least cost.

Paper can be made from a variety of materials—from fabric, wood, and rice to chemical and fiber compounds. The basic papers used in menu design are wood-based with chemical coatings. Recycled papers are also popular; these are waste papers that have been reprocessed into usable material.

Frequently, more than one kind of paper will be needed for the total menu design and depends on the physical format selected. For instance, if a two-panel multipage format has been adopted, two types of paper stocks may be chosen. If the menu format is multi-fold, only one type of paper stock will be needed. Whether one, two, or even three types of paper stock are to be used, there are some important characteristics to look for in paper: strength, finish, brightness, weight, and grade.

Strength means durability. A strong paper will hold up under use. Strength in paper can be determined by the basic material of the paper and the _grain_ of the paper—the direction in which the fibers run in relation to the ways that the paper is folded and cut. Strong paper also tends to have a higher _caliper_ measurement, or bulk thickness.

Grain affects the durability of a menu. The paper will crack and tear if it is folded the wrong way; that is, against the grain. A string-grained paper is stiff and will hold up longer. Moisture, whether in the form of humidity or liquid, can cause paper to expand or contract. If the paper has been folded against the grain, moisture can cause severe buckling. Liquid spills on paper folded *with* the grain, on the other hand, will bead up before becoming absorbed into the paper, allowing time to wipe off the spill.

Finish is the texture of the paper. Finishes can be very smooth, as in coated paper, or have a slightly rough surface, as in antique, eggshell, or vellum paper. *Embossed* paper is made by pressing a raised textured surface, such as a rough pebble texture, or a tweed, linen, or geometric pattern, into the paper. Embossed paper is often suitable for menu covers. *Coated* paper is made by applying a layer of finish to the paper during the paper-making process, ranging from a high-gloss to a dull-coated surface. Coating is different from lamination, which is the bonding of a separate sheet of plastic to a sheet of paper. Menu covers are often laminated. Coated paper is more opaque than uncoated and is an easier surface on which to print. Coated paper is also a good choice for menu covers, as it increases durability and is easy to clean. The coating process seals the paper and decreases absorption of spills and stains. The Hard Rock Cafe menu in Figure 7-24 is designed to replicate a 45 rpm record. The paper used for the record is glossy with raised ridges, simulating the surface texture of this type of record. The surrounding paper has a dull finish for contrast.

The *brightness* of the paper affects the quality of the color reproduction and the readability of the menu. Brightness refers to the amount of reflection that is given off by the final printed menu. Too much reflection can cause glare which, depending on the background color, typeface, and layout, can make the menu difficult to read.

As mentioned earlier, the strength of the paper depends in part upon its caliper measurement, or bulk. Thickness should not be confused with the *weight* of the paper. Paper is manufactured and sold according to *ream weight*—the number of pounds that five hundred sheets (a ream) weigh in a basic size. Two different types of paper can have the same thickness but different weights, depending on the way they were manufactured.

In general, paper prices are calculated on a thousand-sheet basis, equivalent to two reams. Paper might be listed in a catalogue in the following manner: 20 × 26-140M. The numbers 20 × 26 refer to the standard size of this particular cover paper, M is the Roman numeral for one thousand, and 140 is the weight in pounds for two reams. Although listings vary, these identification symbols are basic for standard measurement. The metric system has a slightly different set of symbols.

Grade is the name designated to the paper type, based on how the paper is most commonly used. *Bond* is the thin correspondence paper used by the business community. *Text* paper has a textured surface and is available in a wide range of colors. It is frequently laminated and used for menu covers and insert pages. *Cover* paper, as the name implies, is a heavy stock, available in many different finishes and textures, that is used primarily for menu covers. It is exceptionally well suited for menu use for the following reasons:

1. It is strong enough to hold folds well.
2. The thickness allows it to be *scored* (lightly cut to facilitate folding).

3. The surface is suitable for printing, as it holds ink well.
4. Its strength makes it exceptionally durable.

In addition to the types of paper mentioned above, the menu planner can also consider parchment paper, Bristol board, and a variety of drawing papers. All kinds of paper can be laminated (sealed with plastic) to make them stiffer, more durable, and easier to clean. One rule of thumb for selecting paper for menu use is that the paper used for covers should be approximately twice the thickness of the insert pages.

Color

To design an attractive, interesting menu that highlights specific areas or items requires the creative use of color. More than any other aspect of the physical menu, color affects the customer psychologically. Color can be used to create a mood, establish or reflect a restaurant's image, stimulate a customer's appetite, and promote sales of high-profit items.

The subconscious and conscious effects of color are by now relatively well known. The Lucher Color Test, developed to test the reactions of people to various colors, showed that each color induces a distinctly different reaction. Colors can make a person feel cold, hot, heavy, happy, depressed, romantic, and so on.

Color is any part of the spectrum other than white, which is the absence of color. There are three features by which colors are distinguished: hue, intensity, and value. _Hue_ refers to the particular shade or tint of a color. _Intensity_ is the degree of the color's purity, or whether the color has been softened, muddied, or had many other colors mixed with it. _Value_ refers to the lightness of the color. A pale yellow, for instance, would be lighter in value than a deep purple, but a very intense yellow would be greater in value than a weak lavender. The hue, intensity, and value of a color all affect the impact of the color. A bright green, along with some oranges and yellows, on a menu would be associated with freshness, sunshine, and pleasant citrus and vegetable tastes and have a positive effect on the appetite.

In general, certain colors have definite effects on people. Deep reds and purples convey richness and opulence, while beige, pink, light green, and lavender imply a warm and soothing atmosphere. Different cuisines are associated with particular colors—German foods with woodsy browns and greens; Italian with red, white, and green; Mexican foods with gay fiesta colors; Chinese foods with reds and blacks; and French with yellow and gold, to name a few. Because these associations are already fixed in the customer's mind, the menu designer can take advantage of these color combinations when dealing with international cuisines.

To select colors for the menu, the designer should first establish what the desired impact on the customer is to be. Is color to be used to promote sales or create a mood? If a mood, what kind of mood and which menu items are to be featured? To create a mood, color is used mainly in decoration, trim, background, type, borders, and underlining to accent specific courses or items. After the desired effect has been determined, the menu designer should consider how many colors to use. Color combinations should be selected just as carefully as individual colors, for competing colors will cancel out the psychological effects that might have been achieved by the use of individual colors. As with layout, contrasting but compatible elements can be used.

The use of color in the printing process is based on the theory of *three-color vision;* that is, that all the colors of the rainbow are based on the three primary colors red, yellow, and blue. In four-color separation, the original picture is photographed four times through four different filters—black, red, yellow, and blue. (Although this process is based on the theory of three-color vision, black is nearly always used for clarity and depth.) When the original is photographed through the red screen, for example, all colors but red are screened out: this process is repeated with the three other filters. The various different hues of the final product are made by superimposing the screens upon one another in a technical process, just as the colors of the spectrum are produced from combinations of the three primary colors.

The cost of color reproduction is based on the number of colors that are involved in the complete menu design. All four colors need not be used; there are two-color and three-color separation processes. In addition to color separation, however, there are many other steps in the correcting and proofing process, all of which make color expensive to use in menu design. The number of colors to be used in the menu design, therefore, should be considered in terms of cost as well as visual impact. Color can be effective when used sparingly if incorporated into the design where it will have the greatest impact.

Color is one of the most important elements in menu design and its overall effect on how the customer buys should not be ignored. Technology is continually creating new methods to simplify the printing process with techniques that often make design effects possible which previously would have been extremely expensive to produce. Consult a printer about the many processes available in color printing after your basic design decisions have been made, and plan the final design based on total printing expenditures.

Illustration and Graphic Design

After the general layout and other basic design elements have been decided upon, decorative details are added to the menu to spark customer interest, highlight menu copy and cuisine style, and reflect the interior design of the restaurant. These details are referred to as illustration and graphic design.

Illustrations are drawings or diagrams of particular subjects, whereas graphic design refers more specifically to the decorative patterns, borders, or designs that are used, for instance, to separate sections of the menu. Graphic designs are also drawings and paintings.

There are a few simple rules that apply to the use of design and illustration on the menu. First, the art should reflect the cuisine and interior design of the restaurant. To create an impression of overall consistency, designs, illustrations, borders, and typeface should support the same general theme.

Second, the menu design should appear clean and uncomplicated at first glance. A balanced design is important if the menu is to be easy to read. If some elaborate elements are selected, the remaining elements should be comparatively simple to avoid a cluttered look. Different parts of the menu design should not compete for the reader's attention. In the same way, menu art should not confuse the customer. Too many "busy" illustrations and borders can crowd a menu, distracting the cus-

◆ **FIGURE 7-30.** An evening reception at the Hotel Maurice, Paris in 1808.

tomer from making menu item selections. If the artwork overpowers the food items, customers will most likely overlook many selections and sales will suffer.

Third, be aware of the costs of reproducing designs and illustrations. As long as additional colors will not be needed for the artwork, the cost of printing will not change from the basic costs for color separation (for the rest of the color used in the menu). The cost of layout assembly, however, may increase if the designs are very intricate or involved. Know what the graphic requirements are before obtaining a cost estimate.

Menu designers are always on the lookout for resources for menu illustration material. The paintings in Figures 7-30 and 7-31 are early-nineteenth-century French. Figure 7-30 depicts an evening reception at the Hotel Maurice in Paris in 1808. The painting in Figure 7-31 is an intimate portrait of a dinner party painted by Paul Chabas. Both of these paintings were located in the book *Grand Hotel: The Golden Age of Palace Hotels, An Architectural and Social History,* and are excellent for establishing an elegant, fine dining atmosphere. The pen and ink sketches of fish in Figure 7-32 are similar to those found in clip art books and are good for illustrating full-service restaurant menus.

Complete books and software programs of graphics and illustrations (known as "clip art") are available for general use in limited numbers without copyright. Because of the high quality of the material available, the menu designer can review these sources and identify art work that can be copied by printers, instead of hiring an

◆ **FIGURE 7-31.** An intimate portrait of a dinner party in France during the early nineteenth century by Paul Chabas.

artist to create expensive designs for the menu. Figure 7-33 is an example of clip art illustrating fresh vegetables to support the appetizer and salad menu sections for the Italian cuisine menu from Filippo Ristorante in South Boston. Figure 7-34, another example of clip art used for illustration to which dialogue has been added, is taken from the menu for Rhett's Restaurant in the Opryland Hotel, Nashville, Tennessee. On the appetizer panel the illustration is used for humor. Artwork on the second panel is intended both to create humor and, in the case of the whaling picture, to support the menu section devoted to fish items. These two pages are part of a multi-page book with leatherette covers. The breakfast menu from the same restaurant uses a colorful painting of a hunt scene on a Southern plantation to illustrate the breakfast menu on this oval-shaped menu in Figure 7-35. The surface of the menu is coated with a protective semigloss finish. Although expensive to produce, the quality of both Rhett's menus help to establish the customer's perceived value of the restaurant and establish its position as the main dining room for the hotel.

The menu from the main dining room at the Radisson SAS Scandinavia Hotel in

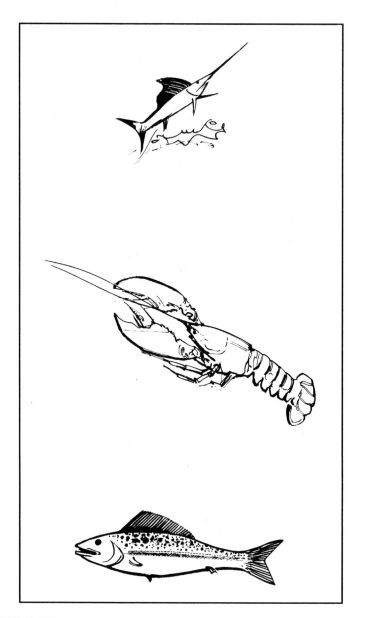

♦ **FIGURE 7-32.** Pen and ink designs for menu illustrations. Reprinted with permission of Ad Art Litho.

Copenhagen, Denmark, shown in Figure 7-36, utilizes original line drawing art to illustrate the menu layout. A photograph of a stream of multi-colored beans, herbs, and spices flows through the menu. The result is a dramatic and organized menu presentation creating customer interest by means of illustration. Musical notation supports the supper club theme of the menu in Figure 7-37. The daily menu in Figure 7-38 uses paper stock printed with illustration, graphics, and a hot red, yellow, and black color scheme selected to support the Mexican theme of the restaurant. This

Antipasti

Focaccia Sulmontina   6.75
Freshly baked rustic country bread topped with the bounty of the garden.

Mozzarella in Carrozza Lucia   7.50
A favorite of many of our patrons. This dish is made with fresh mozzarella and prosciutto in a rich tomato sauce.

Rotolini di Bresaola e Caprino   8.50
Thin slices of the finest cured beef rolled and stuffed with goat cheese.

Quaglia dei Pastori   8.95
Fresh quail stuffed with meat, herbs and spices.

Battello di Gamberi e Capesante   8.95
Fresh shrimp and scallops seasoned with herbs.

Pinzimonio con Salumi   9.50
A visit to the Italian countryside would immediately send the host/hostess off to prepare a welcoming table. This dish is made with fresh and sott'olio vegetables, cheeses and cured meats.

Prosciutto con Melone   9.50
Prosciutto di Parma with fresh melon in season.

Peperonata con Mozzarella   9.50
Fresh roasted red peppers with imported bufala mozzarella.

Lumache Sotto Bosco   9.50
Snails filled with cream of porcini mushrooms, cream of tartufo and herbs and spices.

Focaccia e Battello (priced according to market)
Rustic country bread with fresh shrimp, mussels and scallops seasoned with fresh herbs. This dish could vary depending on the market.

Polenta tra Boston Garden   8.75
Oven-baked polenta with garden fresh herbs and salsicce (sausage) in a sammarzano tomato sauce. Served in a pan hot from the oven.

Insalate

Insalata del Giardino   4.25
A garden fresh salad of crisp romaine lettuce.

Radicchio e Indiva   6.50
Fresh radicchio and endive salad.

Melanzane Sott'Olio   6.50
Eggplant marinated in olive oil.

Peperonata Abruzzese   8.50
The finest imported red peppers are roasted and peeled, lightly seasoned and marinated in extra virgin olive oil.

Rugola, Formaggio e Mele   9.50
Salad of imported rugola, a special variety of cheeses and apple.

Tra Verdi e Pavarotti   10.95
Fresh salad of crisp romaine, endive, and radicchio with tomatoes, onions and sliced broiled chicken with our house salad dressing.

◆ **FIGURE 7-33.**   Graphics illustrate this menu for Filippo's restaurant in Boston, Massachusetts. Reprinted with permission of Filippo Ristorante, North End, Boston.

page is inserted on a panel and changed whenever management wants to feature new and different menu items.

Photography can be an effective decorative medium as well. The page shown in Figure 7-39 is the menu layout for the menu cover shown in Figure 7-17. Photographs for this promotional menu illustrate specific menu copy with fresh, natural, and contemporary images, projecting the general theme and decor of the restaurant. The major problem with this menu is production costs. Photography costs, reproduction, and die cutting charges add up to an expensive menu.

Hot Appetizers

HOT SMOKED PEPPERED SALMON
*With apple—ginger
relish and appropriate condiments.*
$8.25

CREAMY WILD MUSHROOM
AU GRATIN
*Tender sautéed wild
mushrooms with Jack Daniel's Whiskey
sauce and topped with cheese.*
$8.75

COUNTRY STYLE MUSHROOMS
*Large mushroom
caps filled with country sausage from farms
nearby, baked in beef "Au Jus," and
finished with a creamy peppercorn sauce.*
$6.25

PAN FRIED BLACK-EYED PEA CAKE
*Traditional southern
favorite starter with chow-chow
relish, sour cream and salsa.*
$4.75

*What do you mean
you gave away my
reservation?*

*Look away! Look
away! Look away and
I'll eat everything on
your plate!*

From the Waters Nearby

STONES RIVER "FILET & SHRIMP"
*A tender filet of beef
cooked to your liking, complimented by
delicately broiled marinated jumbo shrimp.*
$25.50

CORN ON THE CATFISH
*Corn-fed farm raised
catfish filet rolled in white and yellow
corn meal, and fried to golden perfection.
Served with hush puppies and cole slaw.*
$14.50

CHEROKEE FOREST RAINBOW TROUT
*Rainbow trout sautéed
with mushrooms and bay shrimp, served
with rice and vegetable of the day.*
$15.25

FILET OF RED SNAPPER
*Red snapper from the
Atlantic fishing grounds sautéed in butter
and served with a citrus chive beurre blanc.
Accompanied with rice and vegetable of the day.*
$21.95

NORTH AND SOUTH COMBINATION
*Filet mignon and
one lobster tail cooked to perfection.
Served with drawn butter,
potato and vegetable of the day.*
$36.95

*I don't know.
It looks a little big
for a catfish.*

*Extra! Extra! Yanks
drop Rebs... in fire*

♦ **FIGURE 7-34.**   Illustrations with dialog text are used for humor on this menu from Rhett's at the Opryland Hotel, Nashville, Tennessee. Reprinted with permission of Opryland Hotel, Nashville, Tennessee.

Cover Design

While it may not be possible to tell a book by its cover, a menu cover should reflect the identity of the restaurant as much as possible. A well-designed menu cover should convey the image, style, and cuisine of the restaurant and establish a perceived value on the customer's part for the menu prices. The quality and condition of menu covers also convey the attitude of management toward the restaurant operation.

The image of the restaurant can be portrayed through the menu cover by supporting the mood or theme established by the interior decor. In general, a restaurant's image is directly related to the pricing structure and the type of cuisine offered.

The menu cover for El Patio, shown in Figure 7-40, features a fountain surrounded by gardens. The image that the illustration reflects is sedate yet casual, elegant, but not fussy. Customers would expect a menu featuring light items such as fish and

♦ **FIGURE 7-35.** The breakfast menu from Rhett's is richly illustrated with a colorful scene of Southern plantation life in the nineteenth century. Reprinted with permission of Opryland Hotel, Nashville, Tennessee.

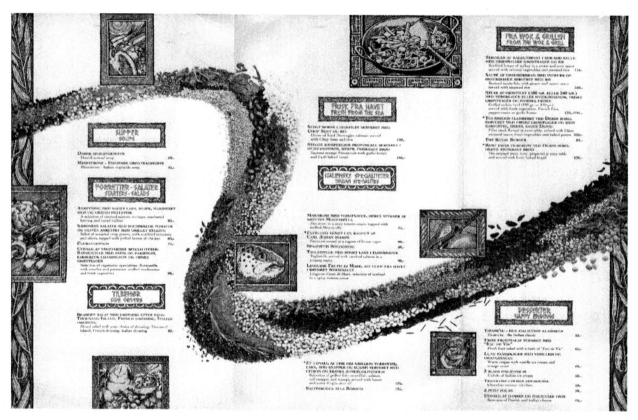

♦ FIGURE 7-36. Original line drawings illustrate the section headings for this Scandinavian menu. A colorful photograph of a mix of beans, herbs, and spices keeps the reader's eye traveling through the menu. Reprinted with permission of Radisson SAS Scandinavia Hotel, Copenhagen, Denmark.

poultry, salads, and perhaps an overall cuisine influence in the medium to higher price range. On the menu cover for Felicity's Dining Room at the Mystic Seaport Inn, shown in Figure 7-41, a watercolor emphasizes the waterside location of the restaurant and primarily seafood menu. The style of the painting expresses a relaxed atmosphere in addition to the well-established identity of the restaurant. Customers would expect a wide range of menu prices as well as menu items.

The menu cover for Harry's Savoy Grill in Wilmington, Delaware in Figure 7-42 conveys sophisticated elegance which supports the fine dining concept of the restaurant. Customers would perceive that the menu offered higher priced items. The use of black as the primary color for the menu cover with light gold highlighting the design and Harry's signature is striking. The illustration of a Japanese samurai warrior on the cover of this Japanese menu in Figure 7-43 establishes that the menu will offer a fine dining atmosphere and Japanese cuisine. The quality of the illustration combined with the heavy cover backed in soft vinyl definitely prepares the customer for a quality foodservice experience.

Figure 7-44 is the menu cover for a smorgasbord restaurant in Helsinki, Finland. The cartoon character establishes a fun and relaxed restaurant atmosphere. The

◆ **FIGURE 7-37.** Musical notation illustrates this menu.

DINNER

APPETIZERS

Cheese and Smoked Turkey Quesadillas
with Green Chile Chutney and Spiced Red Pepper Dip
$7.75

Gazpacho
with White Corn and Blue Potato Chips
$4.75

Ceviche of Halibut
with Citrus and Serrano Chiles
$7.95

Mesclun Salad Greens
with Goat Cheese Tortilla Croutons and
Chipotle-Orange Vinaigrette
$4.75

ENTREES

Achiote and Orange Roasted Chicken
with Chocolate Mole
$18.95

Grilled Snapper
with Green Mole and Red Onion Pico de Gallo
$19.95

Fried Green Tomato, Pozole Poblano and Bean Chile
with Jalepeño and Jack Cheese Corn Bread
(Vegetarian)
$14.95

Carne Asada
Chile Rubbed Strip Steak with Southwest Stuffed
Potatoes and a Tomato Salsa
$23.95

DESSERTS

Cactus Pear and Tequila Sorbet
with Cinnamon Tortilla Strips

◆ **FIGURE 7-38.** A daily menu uses southwestern theme illustrations to support the cuisine theme of the menu. Reprinted with permission of the Delta Chelsea Inn, Toronto, Canada.

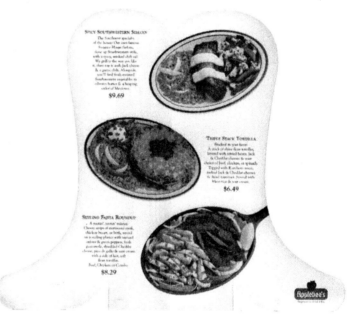

◆ **FIGURE 7-39.** A promotional menu from Applebee's restaurants uses photography to illustrate menu items. Reprinted with permission of Applebee's Restaurants.

♦ **FIGURE 7-40.** The central court fountain at the El Patio restaurant at Caylabne Bay Resort in the Philippines is the focus of this menu cover. Reprinted with permission of Caylabne Bay Resort, Philippines.

♦ **FIGURE 7-41.** The anchor in the foreground of this menu establishes the setting for Felicity's Dining Room at the Mystic Seaport Inn, Mystic, Connecticut. Reprinted with permission.

♦ **FIGURE 7-42.** A sophisticated menu cover in black with gold ink supports the full-service menu at Harry's Savoy Grill in Wilmington, Delaware. Reprinted with permission of Harry's Savoy Grill, Wilmington, Delaware.

◆ **FIGURE 7-43.**   A Japanese samurai warrior illustrates the menu cover for the Benkay restaurant at the Hotel Nikko in Manila, Philippines. Reprinted with permission of Dusit Hotel Nikko, Manila, Philippines.

◆ **FIGURE 7-44.** A cafe and smorgasbord restaurant uses this cartoon illustration for the menu cover of the Stroget restaurant at the Radisson Hotel in Helsinki, Finland. Reprinted with permission of Radisson SAS, Helsinki, Finland.

♦ **FIGURE 7-45.** Menu item ingredients are used to create this colorful photograph for Mama's and Papa's restaurant at the Radisson SAS Scandinavia Hotel in Copenhagen, Denmark. Reprinted with permission of Radisson SAS Scandinavia Hotel, Copenhagen, Denmark.

actual menu is an oversized two-panel with a light moss shade of green as the background color for the illustration. Figure 7-45 is the menu cover for the layout and illustration example seen in Figure 7-36. Raw vegetables, grains, beans, pastas, olives, vegetables, spices, shellfish, herbs, and nuts are used in the design of a bouquet. Without having seen the inside of the menu, customers can assume the

♦ **FIGURE 7-46.** The view of Rockport Harbor, Maine from the dining room of the Sail Loft illustrates this menu cover. Reprinted with permission of the Sail Loft, Rockport, Maine.

use of fresh foods prepared with the wide variety of spices and herbs shown here. This very colorful photograph provides customers with an interesting form of food art that is both informative and interesting.

A painting of Rockport Harbor, Maine is the feature of Figure 7-46, the Sail Loft restaurant menu, mirroring the setting of the restaurant. The design of the restaurant includes a wall of windows looking out onto the picturesque harbor. The customer can expect New England cuisine featuring seafood. The menu cover design from the Sonesta Hotel on Key Biscayne, Florida clearly establishes an oceanside location in Figure 7-47. This restaurant is well known in the Miami area for its seafood buffet. The artwork suggests a full-service restaurant with a relaxed atmosphere.

♦ **FIGURE 7-47.** A colorful painting of dolphins underwater is the menu cover of the dining room at the Sonesta Hotel, Key Biscayne, Florida. Reprinted with permission of Sonesta Beach Resort, Key Biscayne, Florida.

The whimsical cartoon character supported by a large strawberry illustrates the name of the restaurant, Croc'n Berrys, on the menu cover shown in Figure 7-48. A casual, family-style atmosphere is expected with a menu offering a wide range of items, including crocodile.

Each of the menu covers discussed above has been designed to reflect the image of the restaurant for customers before they have reviewed the menu selections

♦ FIGURE 7-48. The cartoon characterization of a crocodile illustrates the menu from J.T.'s Croc'n Berrys. Reprinted with permission of Ad Art Litho.

and prices. Establishing an image means creating expectations for the customer. Customers will have preconceived expectations for both the price and quality of food for their restaurant experience established by the menu cover.

In addition to expressing the image of the restaurant, menu covers help to set the mood or atmosphere for the restaurant. The menu in Figure 7-49 establishes a casual mood and fun atmosphere with a colorful cartoon illustration for the Snapper Bar & Grill in Desdin, Florida. Two contrasting graphic styles for similar theme restaurants reflect different levels of customer expectation and cuisine. The menu cover in

♦ **FIGURE 7-49.**   The Lucky Snapper Grill and Bar in Desdin, Florida sets the mood for an oceanfront casual family restaurant. Reprinted with permission of the Lucky Snapper Bar and Grill, Desdin, Florida.

Figure 7-50 is a sophisticated design incorporating a large menu cover using the contrast of soft pastels against black to establish the mood for a menu offering signature Southwestern dishes. The design for Little Anita's in Figure 7-51 is the menu cover of a multi-panel menu offering standard Mexican cuisine items in the casual relaxed atmosphere of a cantina. The illustration on the cover of the menu for Wittle's restaurant in Figure 7-52 clearly sets a light, casual tone for a hotel coffee shop.

◆ **FIGURE 7-50.** Bolo's Southwestern Grill in Visalia, California uses a desert theme to illustrate their menu cover in black and turquoise blue. Reprinted with permission of Radisson Hotel, Visalia, California.

◆ FIGURE 7-51. Bright red and green is the color scheme of the menu cover for a Mexican cuisine restaurant. Reprinted with permission of Ad Art Litho.

♦ **FIGURE 7-52.** A light, casual tone is established for a hotel coffee shop with this menu cover illustration. Delta Hotels. Reprinted with permission of the Delta Chelsea Inn, Toronto, Canada.

♦ **FIGURE 7-53.**   An entertainment theme is clearly established with the cover of the menu from Harley Davidson Cafe, New York, New York. Reprinted with permission of Harley Davidson Cafe, New York, New York.

Sports-oriented restaurants offer a range of themes. The menu cover in Figure 7-53 for the Harley Davidson Cafe supports the interior design of the restaurant, which features motorcycles, associated accessories, and memorabilia. The bottom of the menu cover is cut out around the wheels and opens vertically. The menu layout can be seen in Figure 7-6. The basketball theme of Michael Jordan's restaurant in Chicago is graphically supported by the menu cover in Figure 7-54, which offers a close-up look at the surface of a basketball, in bright orange and black. Restaurants focused around entertainment themes also provide excellent opportunities for creative menu covers, such as the electric guitar die cut design shown in Figure 7-16a.

♦ **FIGURE 7-54.** The texture of the surface of a basketball is the illustration for the menu cover of Michael Jordan's, The Restaurant in Chicago, Illinois. Reprinted with permission of Michael Jordan's The Restaurant, Chicago, Illinois.

◆ **FIGURE 7-55.** Commander's Palace in New Orleans, Louisiana creates a celebration theme with this specialty menu cover. Reprinted with permission of Commander's Palace, New Orleans, Louisiana.

Although customers approach a restaurant in a variety of moods over which management has no control, every effort should be made to put the customer in a frame of mind to enjoy their experience. An overall ambiance that includes location, music, decor, uniforms, and entertainment will help the customer to participate in the experience rather than just observe it. The menu cover in Figure 7-55 from Commander's Palace in New Orleans asks the customer to join in the party, whether it's Mardi Gras season or not. Although the menu cover may seem a very small part of the total picture, it is an important one. A good menu design reflects management's attitude toward the quality of the operation. The condition of the menu can make just as much of an impression. Customers react to the condition of the menu as a direct reflection of management's sanitation and operations policies. If a menu is covered with stains, spills, and splatters of food, there is a good probability that the kitchen and service areas look the same way. While accidents happen (and soiled menus may be overlooked if an operation is busy), grease and stains on a menu project the image of dirty floors behind the counters and roaches in the kitchen. Menu covers made of coated or laminated papers are easy to keep clean and hold up longer under use, cutting down on replacement costs. Professional, clean, well-constructed menus help to convince customers that management is concerned with every aspect of the operation of the restaurant.

As with every part of the basic design, the cover is an essential element of the total menu. The design concept, mechanical parts and production expenses must be given careful consideration.

CATERING MENU DESIGN

Catering menus require an entirely different overall design concept than do restaurant menus. Catering menus are often presented in a package format combining up to 25 or 30 separate menus as well as theme packages, beverages, and catering policies. Menu presentation techniques can influence which catering service, club, restaurant, or hotel is chosen. In many instances, catering menus are reviewed by customers in their homes or offices without the assistance of a sales representative or other professional who can guide their selection, which often results in decisions being made that are not in the customer's best interests. When these situations occur, the time-consuming process of restructuring menu selections to those that will produce the desired outcomes occurs.

Like restaurant menus, catering menus should be designed so that customers are directed to select the most profitable catering products and services. A key element for sales and marketing—impulse buying—is a major factor in restaurant menu design. Catering menu design, while still offering the opportunity for impulse buying, must approach marketing from a different perspective. Customers have a long-term period in which to compare and discuss menu selections, particularly for social occasions. It often becomes the responsibility of whoever finalizes the meeting and/or event to "upsell" the menu and/or package services. Catering menu design and pricing should allow for the opportunity to upgrade and add on products and services to function arrangements.

The cost of catering menus is a challenge to the overall design process. As customers for catering functions require a number of menu selections to review, reproduction costs are a major budget consideration. In the following section, a variety of catering menu design formats are reviewed and discussed relative to costs and marketing impact. Presentation formats for catering menus include the following design elements:

♦ Presentation covers
♦ Design format

   ◆ Layout
   ◆ Typeface
   ◆ Paper and color
   ◆ Illustration and graphic design
   ◆ Copy

The general rules for these elements remain the same as those discussed for restaurant menus. The application of these elements as applied to catering menu design is reviewed in the following discussion.

Sales Presentation Covers

The design of the overall sales presentation cover for catering menus is dependent on the marketing objectives of the operation. Design formats, as shown in Figure 7-56, can range from a two-panel fold with single page inserts to a multi-page book

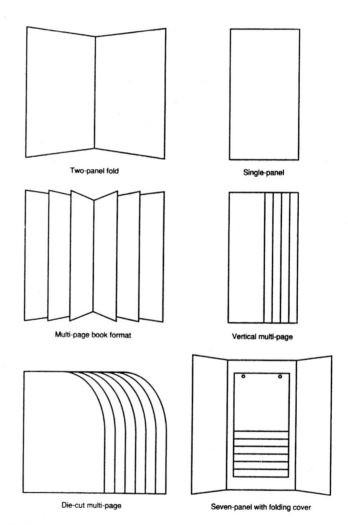

◆ **FIGURE 7-56.** Catering menu presentation designs.

♦ FIGURE 7-56. A two-fold sales presentation cover from Amelia Island Plantation, Amelia Island, Florida. Reprinted with permission of Amelia Island Plantation, Amelia Island, Florida.

format. The sales presentation cover in Figure 7-57 is a two-panel fold that holds a package of single-panel pages. As seen in Figure 7-58, catering policies and information, audio visual equipment rental prices, and additional services are included along with specific menus and packages tailored to suit individual client needs. The cover design theme is repeated on the pocket covers. The cream and light blue design reflects the ocean location. The Plantation brochures are included in the package. Overall design incorporates menu and information pages that are generated from a laser printer using regular office stationery for paper stock.

The sales presentation cover design in Figure 7-59 offers a multi-purpose design appropriate for both social and business functions. An eye-catching contemporary design in soft, muted colors sells the idea of new and original while being simultaneously traditional. The term Catering Concepts can easily include themed functions as well as hospitality and meeting break ideas.

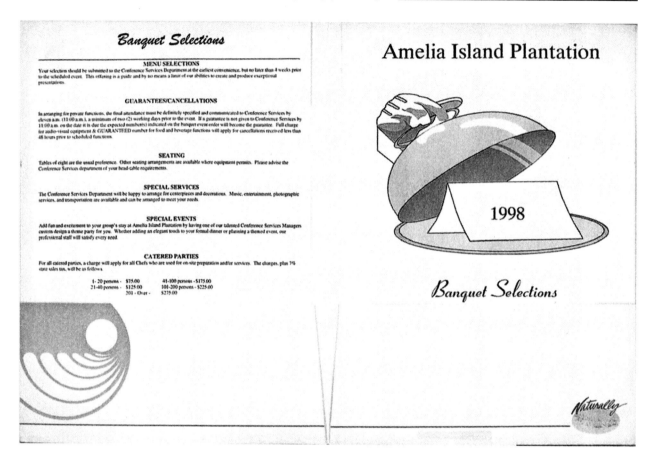

◆ FIGURE 7-58. Catering policies are included in sales presentation folders. Reprinted with permission of Amelia Island Plantation, Amelia Island, Florida.

The stark black presentation of the multi-purpose cover in Figure 7-60 is broken only by the small logo figure in gold placed in the lower left-hand corner. The figure is picked up in the catering stationery shown in Figure 7-61. These single-page menus are created on an as-needed basis with a laser printer and the logo stationary.

A variation for the catering sales presentation of menus and policies is seen in Figure 7-62 from the Garden of the Gods Holiday Inn in Colorado Springs, Colorado. The index design creates a sense of order in the package presentation and facilitates identifying the different areas of information. In the selection featured in Figure 7-62, the menus are referenced to a separate price sheet.

Figure 7-67 from the Radisson Airport Hotel and Conference Center in Columbus, Ohio is a multi-page book format incorporating food photography. The cover, shown in Figure 7-63, utilizes a paper stock with a bronze colored, semi-gloss finish. The 24 design pages include both social and business catering package information. This dramatic presentation, seen in Figure 7-64, establishes quality and perceived value

◆ **FIGURE 7-59.**   Artistic design for catering sales presentation folder. Hyatt Hotels.

◆ **FIGURE 7-60.**   The sales presentation cover from the Portland, Maine, Marriott. Reprinted with permission of Portland Marriott at Sable Oaks.

Lunch Buffet Style

The Italian
Minestrone Soup
Antipasto Salad
Garlic Bread
Penne' Carbonara
Chicken Milanaise
Italian Sausage Sauteed with
Onions, Peppers
Zucchini Capricio
•
Cannolis & Italian Cookies
$17.95 per person

Oriental Express
Hot and Sour Soup
Chicken and Walnuts with Hoisin Sauce
Barbequed Pork with Plum Sauce
Oriental Vegetables & Stir Fried Rice
Egg Rolls served with Sweet & Sour Sauce
•
Fortune Cookies & Orange Custard
$16.95 per person

Mexican Buffet
Mexican Black Bean Soup
"Fill your own Fajitas"
Soft Tortillas
Magic Beef & Chicken
Lettuce, Diced Tomato, Onion, Black Olives,
Shredded Cheese, Guacamole, Mexicali Salsa and Sour Cream
Spanish Rice
Tri-Colored Chips & Salsa
Red Hot Chili Poppers
•
Fruit Flan
$15.95 per person

Prices do not include our 17% Service Charge & the 7% State Tax

PORTLAND **Marriott**
AT SABLE OAKS

200 Sable Oaks Drive ◆ South Portland, Maine 04106 ◆ (207) 871-8000

◆ **FIGURE 7-61.** Illustrated catering stationery to coordinate with sales presentation cover. Reprinted with permission of Portland Marriott at Sable Oaks

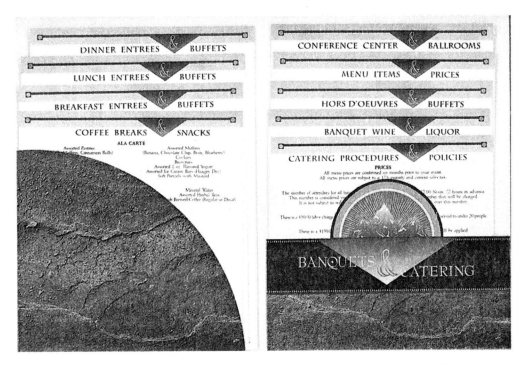

◆ **FIGURE 7-62.** Indexed catering sales presentation. Reprinted with permission of Holiday Inn, Colorado Springs, Colorado.

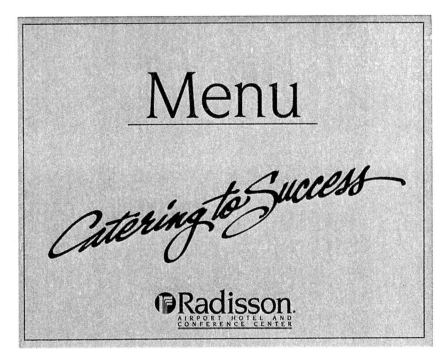

◆ **FIGURE 7-63.** Cover for book format catering sales presentation cover. The Radisson Airport and Conference Center, Columbus, Ohio. Reprinted with permission of Radisson Airport and Conference Center, Columbus, Ohio.

SELECTED DINNERS & ACCOMPANIMENTS

Selected Dinners

Cream of Asparagus Soup, Endive and Bibb Salad with Artichoke Hearts and Dijon Vinaigrette Dressing
Grilled Swordfish Steak with Dill Butter $24.95
Battonets of Fresh Carrots
Buttered Leaf Spinach and Saffron Rice
Soft Rolls, Creamery Butter
Chocolate, Chocolate Torte
Columbian Coffee, Selected Teas

Coquilles St. James (Tender Scallops in a White Wine and Mushroom Sauce, served in a Natural Shell)
Bibb, Radicchio and Endive, Hearts of Palm and Pimiento Hazelnut Dressing
Broiled Strip Sirloin Steak with Sauce Choron $28.95
Bouquetiere of Vegetables
Dinner Rolls and Butter
Brandied Peaches over Almond Ice Cream
Blended Coffee, Imported Teas

Hawaiian Pineapple with Melon Pearls, marinated in Port Wine, Tossed Kentucky Bibb Lettuce with Croutons and Vinaigrette Dressing
Grilled Boneless Breast of Chicken with Sauce Grand Marnier $22.95
Tomato Florentine, Timbale of Wild Rice
Baskets of French Bread and Dinner Rolls
Classic Creme Caramel
Blended Coffee, Selection of Teas

Cantaloupe Crown filled with Tropical Fruits, topped with a Fresh Strawberry
Spinach Salad Mimosa (Sliced Mushrooms, Mandarin Oranges and Shredded Eggs with Warm Bacon Dressing)
Roast Prime Rib of Beef with Creamy Horseradish Sauce $26.95
Bouquetiere of Fresh Vegetables, Anna Potatoes
Bananas Foster Flambe over Vanilla Ice Cream
Blended Coffee, Selection of Teas

Combination Dinners

6 oz. Tenderloin Filet and Shrimp Scampi $28.95
Soup and Tossed Salad with Choice of Dressing
Fresh Vegetable Medley and Potato
Rolls and Butter, Beverage and Dessert

6 oz., Tenderloin Filet and 6 oz. Chicken Breast $24.95
Tossed Salad with Choice of Dressing
Fresh Vegetable Medley and Potato
Rolls and Butter, Beverage and Dessert

6 oz. Tenderloin Filet and 6 oz. Swordfish $26.95
Tossed Salad with Choice of Dressing
Fresh Vegetable Medley and Potato
Rolls and Butter, Beverage and Dessert

6 oz. Prime Rib and 6 oz. Whitefish $24.95
Tossed Salad with Choice of Dressing
Fresh Vegetable Medley and Potato
Rolls and Butter, Beverage and Dessert

Accompaniments

Potato
Redskin Potatoes, Baked Potato, Parsley Buttered Potato, AuGratin, Scalloped, Dutchess, Double-Baked Potato
Add an Additional $1.00 per person

Vegetables
Baby Glazed Carrots, Green Beans, California Medley, Squash, Zucchini, Red Peppers, Sweet Peas and Mushrooms

Desserts

New York Style Cheesecake with Fresh Strawberries $3.25
Strawberry Amaretto Torte Cake $2.95
Chocolate Truffle Mousse Cake $2.95
Carrot Cake $2.50
Swiss Chocolate Mousse $2.25
Selection of Ice Creams $1.95
Bananas Foster $3.50
Selection of Freshly Baked Pies $2.25
Cherries Jubilee $3.50
Key Lime Pie $3.25
German Chocolate Torte $2.95
Baked Alaska $3.50
Plantation Pecan Pie $2.95
Double Chocolate Cake $2.50
Chef's Cobbler $2.00
Chocolate Oblivion Torte on a Raspberry Cloud $3.50

The above items are suggestions and recommendations. Our catering staff will be happy to tailor a menu to your specific needs.

All Banquet Functions will be Charged a 17% Service Charge and Applicable Taxes.

15

♦ **FIGURE 7-64.** Color photography illustrates catering menu selections. Reprinted with permission of Radisson Airport and Conference Center, Columbus, Ohio.

for the customer. The major challenge with this concept is that price and menu content changes cannot be made without reprinting the entire piece. Results from sales with this design will determine the feasibility of its continued use.

The three-panel fold-out wedding sales presentation cover in Figure 7-65 incorporates a die cut outline of the wedding cake on the front panel with an additional

♦ **FIGURE 7-65.** Three-panel fold wedding sales presentation folder and inserts. Providence Marriott, Providence, Rhode Island.

◆ **FIGURE 7-65.**   *(Continued)*

cut-out outline on the inside pocket. Single-page menus and wedding packages with index headings are inserted into the side pocket. The wedding presentation folder measures $6^1/_4'' \times 9\ ^1/_4''$.

Typeface and Illustration

Typeface and illustration are important considerations in catering menu design. As catering menus generally include a lot of text material, it is important that customers be able to read individual menu listings. Figure 7-66 is a good example of a catering menu that takes maximum advantage of both layout and typeface to create a legible menu listing. While the same typeface is used for both pages, section headings and menu item names stand out in boldface type from the descriptive copy. Additional per person charges are also highlighted to make sure that the customer is aware of extra charges. The layout places the individual menu items and sections to take advantage of the effect of lightening created by the black ink against a stark white background. A design incorporating the outlines of the Lincoln Memorial, the Washington Monument, and the Capitol Building has been preprinted in light blue on the paper stock along with the border and Washington Marriott logo.

The menu page in Figure 7-67 utilizes two typeface styles as well as boldface print to present the breakfast buffet menu selections. As in Figure 7-68, information that is important for the customer to be aware of is printed in the most legible typeface. Figure 7-68 shows a menu from a catering program that focuses on highlighting both

♦ **FIGURE 7-66.** Catering menu layout from the Washington Marriott, Washington, D.C. Reprinted with permission of Washington Marriott, Washington, D.C.

regional foods and the history of the City of Savannah, Georgia. An illustration of a statue of the Marquis de Lafayette is in the upper left-hand corner. Descriptive copy introduces regional Southern food items such as native oysters, roast duck, and watermelon. This menu series was preprinted on light blue paper and printed in red ink as opposed to off-white paper stock and black ink for the balance of the catering menu pages.

OVERVIEW

This chapter has covered a wide range of issues related to the overall design of the restaurant and catering menus. When designing a menu, it is important to keep all of the major design considerations in mind as you deal with each one individually. When the total design package is assembled, use the following questions as a guideline to help in making value judgments about the quality of the design and to edit and revise the menu.

FIGURE 7-67. The typeface styles are used to present catering breakfast selections. Amelia Island Plantation, Amelia Island, Florida. Reprinted with permission of Amelia Island Plantation, Amelia Island, Florida.

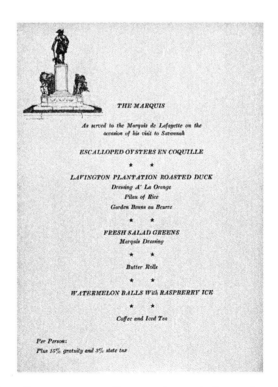

FIGURE 7-68. Illustration reinforces a historical catering menu theme.

Review Guideline Questions

♦ Does the menu design reflect the image that you want to convey to your customers?

♦ Is the menu easy to read?

♦ Does the overall design and layout provide the customer with the ability to make a reasonably quick decision as to what menu items to order?

Take the time to review, edit, and revise the menu material, then re-evaluate menu items and your final menu listing for changes if necessary. Encourage other members of your organization to review the menu and offer their comments. Very often, it is difficult for the menu designer and management to see problems that others outside of the development process can easily spot.

CHAPTER 8

Marketing with Copy

Copy

Menu copy are the statements used to sell menu item selections, promote the restaurant, and educate customers about the services that the restaurant offers. The three general types of menu copy used as marketing tools for menu design are:

♦ merchandising copy
♦ accent copy
♦ descriptive copy

Once management has determined what menu items will be offered and what the prices are to be, the menu designer should focus on marketing these items through the use of menu copy. *Menu copy* is the written text on the menu used to sell individual menu selections, promote the restaurant, and inform the customer of the complete range of services that the restaurant offers. Copy is the principle way of marketing the restaurant product. As noted previously, menu copy is divided into three categories: merchandising, accent, and descriptive copy.

Merchandising Copy

Merchandising copy refers to the written passages on a menu that promote the restaurant. The form of copy provides such basic information as the name, address, telephone, and fax numbers of the restaurant as well as any services that the restaurant may provide, including credit card acceptance, private rooms, take-out and catering facilities, entertainment, and special events. Merchandising copy can also be used to feature information related to the restaurant in some way, such as local historical events, maps and stories about local characters, the history of the restaurant or of the cuisine—anything that might amuse or interest guests.

Some merchandising copy is always necessary. For private parties and banquet service, when menus will be printed for an occasion, the only merchandising copy necessary is the name of the restaurant, location, and telephone/fax numbers. These menus are important marketing tools as they go beyond your doors and will be used as reference by people attending the function. Listing the restaurant's name, address, and telephone numbers provides a way for future business to contact you. Some operators include their e-mail address and/or web site locator.

The decision as to how much merchandising copy to use depends on the management, style, location, and cuisine of the restaurant, as well as the general format of the menu cover and availability of space on the layout. The merchandising copy on the back panels of the Cracker Barrel restaurant menu (Figure 8-1) in addition to a children's menu selection, both describes some of the signature dishes and the cooking techniques applied to their preparation and promotes mail order sales of prepackaged foods. A note at the bottom of the center panel encourages customers to take the menu with them. This example of food item sales reflects the growing trend in the restaurant industry to create an additional avenue of revenues and profit through the sales of in-house merchandise. From cookbooks, as seen on Commander's Palace menu in Figure 8-2, to signature food items, to the on-site merchandise shops such as featured in the restaurants Hard Rock Cafe, Michael

♦ **FIGURE 8-1.** The back panel of the Cracker Barrel restaurants menu is used for extensive merchandising copy. Reprinted with permission of Cracker Barrel Restaurants.

Jordan's, and Harley Davidson, more and more foodservice operators are taking advantage of the captured customer audience waiting for tables for up to 45 minutes or an hour. The Cracker Barrel restaurants have developed a separate retail outlet called Old Country Store that flows directly into the restaurant. Using almost as much floor space as the restaurant itself, the store sells everything from penny candy to rocking chairs. The store brochure is shown in Figure 8-3.

The menu in Figure 8-4 is historical background information for the menu cover featured in Figure 7-43 from the Japanese restaurant Benkay at the Dulsit Hotel Niko, Manila. Copy of this type keeps customers occupied while waiting for servers and provides conversation between guests. The menu cover in Figure 8-5 for the Bermuda Triangle Bar & Grill features merchandising copy to support the menu graphics of a map area including the Bermuda Triangle area. The menu copy tells the story of the area with the purpose of creating customer interest and conversation.

The ultimate combination of merchandising copy and menu layout is the newsletter format featured in a growing number of restaurants. The two examples shown here in Figures 8-6 and 8-8 provide a contrast of casual, family-style theme restaurant operations. Figures 8-6 and 8-7 are from The Roadhouse, a steakhouse with a country western theme in Rehoboth, Delaware. In this format, the menu takes up most of the second and third pages of a four-page newsletter. The front page offers a history of the owner, Country Dan. A Southern country restaurant, Applewood Farmhouse in Sevierville, Tennessee uses a newsletter format, shown in Figure 8-8, to feature apple cider and to promote one of their signature items, home-baked breads. A large,

♦ **FIGURE 8-2.** The Commander's Palace menu in New Orleans promotes the sale of the restaurant's cookbook. Reprinted with permission of Commander's Palace, New Orleans, Louisiana.

The charm of an old country store.

Stepping into a Cracker Barrel Old Country Store is like traveling back to an era when stopping on the road for a meal was special. Out front there's a welcoming front porch filled with cozy rocking chairs. Pass through our wooden doors and walk into a real country store and restaurant. At every turn, there's something new and old to behold: hand-blown glassware and cast iron cookware; aromatic smoked meats; old-fashioned crockery from your grandmother's table; handcrafted figurines; classic children's toys; and genuine antiques and memorabilia. hanging from the ceiling. It's all part of the charm of the Cracker Barrel.

Open Sunday-Thursday
6 a.m. to 10 p.m.

Friday and Saturday
6 a.m. to 11 p.m.

Cracker Barrel Old Country Store, Inc. is a publicly held, non-franchised corporation, whose common stock is traded in the over-the-counter market under the NASDAQ symbol CBRL (abbreviation CdrBd). For a copy of our latest annual report, write to: Cracker Barrel Old Country Store, Inc., P.O. Box 787, Lebanon, Tennessee 37088-0787.

We gladly accept personal checks as well as

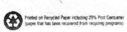

Printed on Recycled Paper including 20% Post Consumer (paper that has been recovered from recycling programs)

Guide to good country cookin'.

We invite you to stop at the Cracker Barrel no matter where your travels might take you.

◆ **FIGURE 8-3.** A brochure for Cracker Barrel restaurants promotes the partner business, the Old Country Store. Reprinted with permission of Cracker Barrel Restaurants.

retail store adjacent to the restaurant sells fruits, vegetables, cider, jams, preserves, and bakery goods. Some operators mail newletters such as these two examples to their customer base seasonally to keep customers informed of new menu items, special promotions, and holiday events.

Suggested topics for menu merchandising copy are:

◆ History of area, historical figures, or buildings
◆ History and stories about the restaurant family organization
◆ A discussion of the restaurant's featured cuisine
◆ Story of celebrities
◆ Trivia questions that tie into the theme of the restaurant
◆ Activities that guests can participate in such as crossword puzzles or fill-in-the blank questions
◆ Local interest stories
◆ Merchandise sales information

The Legend from Whence We Come

"Once on Gojoh Bridge in Kyoto, a giant of a man named Benkay..." So begins a tuneful old song of Japan that tells of an encounter one moonlit night about 800 years ago towards the end of the Heian Period, between the valiant young noble, Ushiwakamaru, and the mighty warrior-monk, Benkay.

A dramatic martial contest ensued on the bridge. The youth's superb agility and skills proved more than a match for the other's strength and size, Benkay – who had never before known defeat – knew then that he had now met his master. Thenceforth, the monk devoted his life in service to the young lord, who later became known as Yoshitsune of the Minamoto Clan, a noble hero, but doomed by the events of time to a tragic end.

To the very end, through all travail and trials, Benkay persevered and served him. The complete devotion and dedication of Benkay is a tale that is still told today in epic drama and theater. Eight centuries have passed since that night on Gojoh Bridge, but the legend of Benkay lives on – in Kabuki theater, Noh drama and films.

The Benkay Restaurant derived its name from this legend, and the monk's dedication is our inspiration. As loyally as he had served Yoshitsune, so we wish to devote ourselves to serving you with the cuisine of Japan. And what could be more fitting than doing so through the special, noted cuisine of Kansai, the area in western Japan where Kyoto lies – with a dedication that even Benkay himself would approve.

◆ **FIGURE 8-4.** Historical legends are the basis of narrative copy on the back of this Japanese menu. Reprinted with permission of Dusit Hotel Nikko, Manila, Philippines.

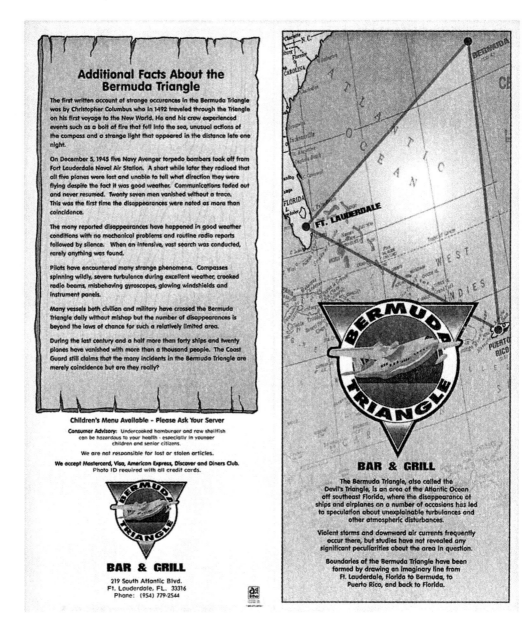

◆ FIGURE 8-5.   Merchandising copy supports the menu graphics for the Bermuda Triangle and Grill, Ft. Lauderdale, Florida. Reprinted with permission of Ad Art Litho.

ROADHOUSE™
J·A·W·B·O·N·E

Midway Shopping Center · (302) 645-6273
1572 Highway One, Rehoboth Beach

Pike Creek Shopping Center (302) 994-6676
4732 Limestone Rd., Pike Creek

STEAK JOINT™

A restaurant serving Southern Delaware, Pike Creek, and the best steaks in town

The Legend Of Country Dan

A Rompin', Stompin' Hombre From North O' The Rio Grande

Nobody recollects exactly where Country Dan came from; some say from down Laredo way, one jump ahead of a posse peeved 'bout a shootout over some senorita. Others swear he drifted into town from the Yukon after settlin' a gold claim with a little help from the law firm of Smith & Wesson. One Milton feller even gave his solemn oath he seen Dan unloadin' off a paddlewheeler in the Broadkill River after makin' the skipper bring him up from Natchez by way of the Horn; seems the captain lost his boat to Dan after a long night of three-card monte.

But everybody remembers clear as branch water the day that Dan first cast his big shadow over the Roadhouse Steak Joint and Saloon (and several surrounding acres).

It was real quiet that lazy afternoon in May — too quiet. The barkeep was polishin' up some glasses and swappin' lies with the few Lewes hombres who had wandered in to wash the seawater off their forked tongues. All of a sudden, the doors bust open and

in runs little Elwood Martin, Earl's boy from out by Rabbit's Ferry. He kinda flails up to the bar, elbows and knees thrashin' like a loose-paddle windmill in a nor'easter.

"Gimme a double!" he sorta croaks at the bartender, all outa breath and with this wild look of horror and amazement in his eyes, like he just see'd a ghost or a Rehoboth boy doin' an honest day's work.

Well, ol' Pete, he sets his glasses down and whips that kid up a Roy Rogers that woulda quenched Trigger's thirst, and tells Elwood to go set down at a table where he belongs.

As soon as he's done gulpin' his drink down, and singin' the first verse of "Mary Had a Little Lamb" in one long belch, he looks 'round the room with that deer-in-the-headlights kind of look and barks out them words that no one who was there that day will ever forget ('cept maybe Lem, who doesn't remember nothin')—"Country Dan's comin' to town!"

Well, you never seen such a bunch of Lewes boys move so fast in your life, 'less maybe at the Georgetown Oyster Eat when somebody hollered "Last one out gotta dance with a woman!"

When the dust settled, ol' Pete, who don't move real quick, figgered he better get his butt out to the tall corn, too, and he started amblin' out from behind the bar. Unfortunately for him, as he wheeled around he snagged the end of his bolo necktie on one of Bonehead's horns and before he had got goin' it yanked him off his feet and he hit his head on a jug of

simple syrup.

He don't know how long he was out, but when he come to he could feel the whole buildin' rumblin', like when he woke up that mornin' in his room at the Belhaven Hotel during the Great Storm of '62. He scrambled for the door, but when he threw it open he could see nothin' but a swirlin' cloud of dust with leaves and trash and some porch furniture flyin' around in it. Pete reckoned it was too late by then to make his break, so he better get back inside and just try to act like nothin' unusual was goin' on and maybe this guy wouldn't bother him.

As soon as he gets back behind the bar, there's a sound sorta like when a C-5A comes in right over top of you on 113 up to Dover, 'cept loud. Next, there's this huge "Whoooosh!" and the doors blast right off their hinges, followed by these rollin' billows of green smoke. There this monstrous, shadowy form begins to take shape in the doorway, and soon just fills it up completely.

By now, poor ol' Pete is just cussin' the day they ever hung up that infernal Bonehead skull on the wall; it appeared it was gonna hang him just as sure as a gallows.

As that putrid fog started to clear a little, Pete started gettin' a gander at what he was up against. This mammoth shaggy head was duckin' through the doorway, like a buffalo bargin' into a prairie dog hole. To squeeze through the door, he had to take off his hat, which appeared to have been made from a wagon wheel and the top half of an Indian teepee. His shirt was from the cover of a Conestoga wagon; it still had "The Donner Party" printed across the back of it. His chaps were made from two whole mountain lion hides. The tails were tied up like suspenders, and the two mighty jaws clenched to

COUNTRY DAN HISSELF

hold his socks up. His moccasins were a pair of war canoes, each with the warrior still in it. "Give me a drink!" he roared, "and make it snappy, 'cause I'm in a real foul mood. I just lost me woman today!"

Well, you can imagine the fright this give poor Pete. He figgers his best chance to live to see another mornin' is to humor this feller and maybe offer him a little bartenderly sympathy. He mixes him up a hogshead full of Wild Turkey and 151 rum, rolls it over to him.

"Sorry to hear about your gal. Some dirty varmint jump your claim?" asks Pete, tryin' to

sound as much the rough ol' cob as possible.

"Nah," growls Country Dan. "Me an' Gertie was a-ridin' long on Harley, me wooly mastadon, when I tried to jump a crick down below Fenwick Island. She goes flyin' off, an' lands on her butt in the marsh. When I goes back to get her, she'd sunk down into the mud an' that's nothin' but this crater a-fillin' up with water."

And it happened just like he told it, down at a place they now call Big Assawoman Bay.

◆ **FIGURE 8-6.** A newspaper format gives background for The Roadhouse in Rehoboth Beach, Delaware. Reprinted with permission of Roadsters Restaurant, Lewes, Delaware.

Accent Copy

Accent copy is used to spark interest in a course or individual menu selection through the creative use of names and headings. For instance, the featured cuisine of Tarpey's restaurant is seafood, with beef and poultry included for variety. Accent headings to introduce courses use terms associated with the sea such as "The Sand Bar," "Tommy's Trident," and "Cameo Cargo." Continuing the theme of the restaurant, accent copy is applied to develop distinctive names for entrees and other foods: "Cabin Boy Clams on the Half Shell," "Stowaway Shrimp," and "Bud's Bermuda Triangle."

Accent copy is also used for merchandising when appropriate. There are no rules other than to keep it interesting, clean, and simple, and to avoid confusing the customer. The menu in Figure 8-9 is an excellent example of the use of accent copy for both course headings and menu item names. This casual restaurant menu uses words related to the beach and ocean to match illustrations, with the overall objective of creating an amusing and lighthearted menu. "Wet Your Appetite," "Tossed About," "Beach Sand-Wiches," "Shoreline Creations," and "Coral Classics" take the place of the traditional appetizer, salad, sandwich, specials, and main course headings. Menu

◆ FIGURE 8-8. The Applewood Farmhouse Retaurant in Sevierville, Tennessee uses a newsletter format to promote take-home food items and provide a walk-away copy of the menu for guests of both the store and restaurant. Reprinted with permission of Applewood Farmhouse Restaurant, Sevierville, Tennessee.

items become "Palm Beacher," "First Mate's First Rate Burger," "Gulf Stream Sandwich," and "The Caribe." While the menu in Figure 8-10 uses traditional course headings, it uses unusual and cuisine-oriented menu ingredients in the item name to create customer interest such as "Achiote and Orange Roasted Chicken," Fried Green Tomato, Pozele Poblano and Bean Chile," and "Cactus Pear and Tequila Sorbet."

Descriptive Copy

As accent copy often provides no explanation of what the individual course or dish may be, *descriptive copy* is frequently added to introduce the item to the customer.

♦ FIGURE 8-9. Accent copy is used for course headings and menu item names in this menu from the Walt Disney World Dolphin, Orlando, Florida. Reprinted with permission of Walt Disney.

As the term implies, descriptive copy describes the item, featuring any quality about the dish that might help increase sales, such as the main ingredient, interesting secondary ingredients, and method of preparation. Although some menu items, such as whole Maine lobster, filet mignon, and prime ribs of beef, are self-explanatory and do not need descriptive copy, a tantalizing, well-written description can increase

DINNER

APPETIZERS

Cheese and Smoked Turkey Quesadillas
with Green Chile Chutney and Spiced Red Pepper Dip
$7.75

Gazpacho
with White Corn and Blue Potato Chips
$4.75

Ceviche of Halibut
with Citrus and Serrano Chiles
$7.95

Mesclun Salad Greens
with Goat Cheese Tortilla Croutons and
Chipotle-Orange Vinaigrette
$4.75

ENTREES

Achiote and Orange Roasted Chicken
with Chocolate Mole
$18.95

Grilled Snapper
with Green Mole and Red Onion Pico de Gallo
$19.95

Fried Green Tomato, Pozole Poblano and Bean Chile
with Jalepeño and Jack Cheese Corn Bread
(Vegetarian)
$14.95

Carne Asada
Chile Rubbed Strip Steak with Southwest Stuffed
Potatoes and a Tomato Salsa
$23.95

DESSERTS

Cactus Pear and Tequila Sorbet
with Cinnamon Tortilla Strips
$6.95

♦ **FIGURE 8-10.** Accent copy using specialty cuisine food words creates a colorful description of menu items.

sales of other items. However, overselling a menu item may actually hinder sales. The menu designer must develop the ability to pinpoint which items require descriptive copy and which do not.

The basic guidelines for writing descriptive copy are as follows:

Be simple. Describe the item clearly and cleanly. Use as few words as possible, making each one count. Keep sentences short so that customers can read through the menu quickly.

Keep grammatical style consistent. While the rules of grammar state that only proper names and the first letter of a sentence should be capitalized, menu planners often choose to capitalize items so that they will stand out. Either choice of style is fine, as long as it is applied consistently throughout the menu. If one ingredient is capitalized, for instance, all ingredients should be capitalized. Menus that are consistent in presentation are easier to read and help the customer decide more quickly. In addition, consistency helps to establish an impression of thoroughness and exactness. Descriptive copy is one of the elements that customers judge in their impression of the overall quality of the service, food, and atmosphere of the restaurant.

Use food-associated words. Many menu designers and planners, in search of newer and greater ways to describe food and impress the customer, use terms such as majestic, embraced, sensuous, or designed. These words have nothing to do with the ingredients of the dish or method of preparation and should be avoided.

Avoid the use of superlatives. Exaggeration—using terms such as mounds, drenched, smothered, to perfection, supreme, and at its best—can turn off the customer. Today, the national trend toward eating smaller portions for health and dietary reasons is becoming widespread. A patron considering "a mound of sauerkraut, smothered with melted cheese and topped with Russian dressing supreme," for example, might feel uncomfortably full before ordering the item!

Over-description can turn a simple menu item into a confusion of sauces and ingredients. Descriptive copy is not a recipe. Include only those ingredients that will sell the item. The following menu copy from Figure 8-11 is a good example of using the ingredients to explain a menu item that might not be familiar to most customers:

> **Portabella Pizza**
> corn meal crusted portabella mushroom sauteed in
> olive oil topped with fontina cheese, a roasted garlic
> tomato coulis, grilled onions and prosciutto ham.

The menu in Figure 8-11 consistently presents good descriptive copy for menu items. The menu in Figure 8-12 from Cascades restaurant highlights food preparation techniques in the descriptive copy, creating additional customer interest. "Marinated, steamed and roasted" describes roast duckling, while "Seared, basted and roasted" describes a Hunan style veal chop. The customer's impression of freshly prepared food items is reinforced by cooking terms. Figure 8-13, from the Nag's Head Inn in the village of Usk, Wales, uses minimal but humerous descriptive copy. "Raised and shot within 10 miles of Usk" certainly implies farm raised and fresh killed. "One of the hardest birds to shoot," however, probably helps to justify the price.

A variety of copy uses are seen in Figures 8-14 through 8-18. The descriptions for mussels casino, grilled chicken salad, and salmon salad take a few simple ingredients and enhance them with food-associated words to create an appetizing text that helps the customer to envision a fragrant, steaming bowl of mussels served with

STARTERS

HOT CRABMEAT AND ARTICHOKE DIP
with crisp sourdough croutons in a fresh herb and cream cheese sauce $4.95

FRIED RAVIOLI
cheese ravioli breaded with locatelli cheese and fresh Italian parsley served with a roasted red pepper sauce $4.50

DRUNKEN CLAMS
steamed little neck clams in a garlic broth with ginger, red pepper and onion with shredded lettuce and scallions $6.95

PORTABELLA PIZZA
corn meal crusted portabella mushroom sautéed in olive oil topped with fontina cheese, a roasted garlic tomato coulis, grilled onions and prosciutto ham $6.95

CLAMS CASINO
fresh topneck clams baked to perfection with casino butter and bacon $5.95

HARRY'S SHRIMP COCKTAIL
jumbo gulf shrimp served with cocktail sauce and lemon wedge $7.95

STUFFED MUSHROOMS IMPERIAL
large Kennett Square mushroom caps stuffed with lump crabmeat imperial $6.95

JUMBO BUFFALO CHICKEN WINGS
served with bleu cheese dressing and celery sticks served mild or hot $4.75

TEMPURA SHRIMP
jumbo gulf shrimp battered and fried until golden brown served with cellophane noodles and sweet and sour sauce $6.95

BASKET OF HOMEMADE ONION RINGS
thick cut and freshly breaded $3.95

BASKET OF HOMEMADE YUKON GOLD POTATO CHIPS
with bleu cheese dressing $3.95

SOUPS

BERMUDA FISH CHOWDER
the soup of Bermuda served with original Outerbridges sherry pepper sauce $2.25

HARRY'S SOON-TO-BE-FAMOUS VEGETABLE SOUP
made with a rich stock and the meat from our award winning prime rib with pearl barley and green peas $2.95

THREE ONION SOUP AU GRATIN
scallions, Bermuda red and Spanish onions baked with provolone cheese $3.95

HARRY'S SOUP SAMPLER
enjoy a sampling of onion soup, Bermuda fish chowder and soup du jour $4.25

SALADS

All dinner entrees include a house salad with choice of dressing. Please feel free to substitute the Harry's or Caesar salad for one dollar.

HARRY'S SALAD
mixed greens, ripe olives, red onion, pepperoncini peppers and feta cheese with a fresh herb vinaigrette dressing $4.95

CAESAR SALAD
freshly grated locatelli cheese and seasoned croutons enhance this famous salad $4.30

HOUSE SALAD
romaine and radicchio lettuce with cherry tomatoes and choice of dressing $2.95

DRESSINGS:
Creamy Garlic Herb Vinaigrette, Harry's Vinaigrette, Ranch, Honey Mustard, Maytag Blue, Thousand Island, Cucumber Yogurt LoCalorie, Pecorino Vinaigrette

ENTREE SELECTIONS

All entrees include a house salad with choice of dressing. Please feel free to substitute the Harry's or Caesar salad for a dollar.

HARRY'S NEW YORK SIRLOIN STRIP STEAK
served with choice of French fries or baked Idaho potato and vegetable du jour
12 oz. Iowa Corn Fed Beef $16.95
16 oz. Iowa Corn Fed Beef with cabernet butter and fried onion rings $19.95

GRILLED LONDON BROIL WITH BÉARNAISE SAUCE
Top Sirloin cut of Iowa beef served over watercress with broiled tomato and garlic "smashed" potatoes $13.95

CHAR-BROILED FILET MIGNON
7 ounce filet served on a crouton with Madeira demi-glace julienne carrots and potato du jour $15.95

GRILLED TOP SIRLOIN OF LAMB
served with roasted garlic, herb roasted tomato, tortellinneed potato and port wine sauce $15.95

GRILLED PORK CHOPS WITH JACK DANIEL'S APPLE PAN GRAVY
center cut pork chops served with "smashed" potatoes, vegetable du jour and spiced pecans $13.95

GRILLED BABY BACK RIBS
slowly roasted lean back Iowa pork ribs served with Harry's barbeque sauce, cole slaw and French fries
Half Rack $11.95   Full Rack $15.95

SAN FRANCISCO SEAFOOD STEW
a house favorite, shrimp, scallops, fish, clams, and mussels in a light tomato, garlic and saffron broth served with rustic croutons $18.95

PAN SEARED ATLANTIC SALMON FILET
wild mushroom ragôut with roasted garlic and fresh thyme, served with orzo and lemon beurre blanc $15.95

GRILLED BREAST OF CHICKEN WITH PORTABELLA MUSHROOM
balsamic vinaigrette served with polenta and seasoned spinach $12.95

CAJUN GRILLED CHICKEN BREAST
with smoked shrimp and andouille sausage gravy served with garlic "smashed" potatoes and vegetable du jour $13.95

CRAB CAKES REMOULADE
"Sweet Fingers" Susie's lump crab cakes deep fried to peanut oil and served with shredded lettuce, spicy corn and red pepper relish, French fries and New Orleans remoulade sauce $16.95

PASTA

All pasta entrees include a house salad with choice of dressing. Please feel free to substitute the Harry's or Caesar salad for one dollar.

SAUTÉED CHICKEN SAVOY
roasted peppers, olives and artichoke hearts. Our signature dish with an addition of wild mushrooms served over linguini
Entrée $12.95   Appetizer $6.95

NANTUCKET SEA SCALLOPS WITH SHRIMP AND SALMON in a fresh basil cream sauce over red pepper linguini
Entrée $14.95   Appetizer $8.95

ANGEL HAIR PASTA WITH RAPINI
sundried and fresh tomatoes, toasted pine nuts in a garlic herb butter with pecorino romano cheese
Entrée $12.95   Appetizer $6.95

ENTREE SALADS

WARM SIRLOIN SALAD
served over romaine and radicchio lettuce with Bermuda onion, julienne peppers, cornichon pickles and char-broiled sirloin seasoned with provençale herbs and Dijon vinaigrette dressing $7.95

GRILLED CHICKEN CAESAR
a large portion of Caesar salad with julienne strips of freshly grilled chicken $7.95

SANDWICHES

served with Yukon Gold potato chips and Chessen pickle spear
Side order of French Fries $1.00
Basket of Fries $2.95

OPEN FACE PRIME RIB
Harry's famous rib on sourdough bread with creamed horseradish sauce and French fries $8.95

CRAB CAKE SANDWICH
"Sweet Fingers" Susie's lump crab cake served on a kaiser roll with tartar sauce and lemon wedge $8.95

CHAR-BROILED SIRLOIN HAMBURGER
8 ounces of ground sirloin, with your choice of American, Swiss or Cheddar cheese on a kaiser roll $5.95

GRILLED MARINATED PORTABELLA MUSHROOM SANDWICH
portabella mushroom, grilled onion, plum tomato and provolone cheese on a kaiser roll $7.95

COBB SANDWICH
smoked turkey on homebaked sourdough bread with crispy bacon, tomatoes, avocado and lettuce with bleu cheese dressing $6.95

TURKEY BURGER
fresh ground turkey, charcoal grilled, with Monterey pepper jack cheese, tomatoes and guacamole on a kaiser roll $6.95

SIDE ORDERS

Basket of French fries $2.95
Basket of homemade onion rings $3.95
Sautéed fresh mushrooms $2.95
Basket of homemade potato chips with bleu cheese dressing $3.95

BEVERAGES

Espresso $2.00
Cappuccino $2.75
China Mist Iced Tea $1.50
Perrier $1.75
Evian $1.95
Orangina $2.00
Coca-cola, Diet Coke, Gingerale, Sprite $1.50
O'Douls $2.75

PLEASE INQUIRE ABOUT OUR CATERING AND PRIVATE PARTY FACILITIES.

SMOKING PERMITTED IN THE GRILL ROOM ONLY. WE ARE A CIGAR FRIENDLY RESTAURANT.

A 17% GRATUITY MAY BE ADDED FOR PARTIES OF 8 OR MORE.

PLEASE JOIN US FOR SUNDAY BRUNCH.

WE WELCOME VEGETARIAN AND LOW FAT REQUESTS.

◆ **FIGURE 8-11.** The menu from Harry's Savoy Grill uses good descriptive copy for menu items. Reprinted with permission of Harry's Savoy Grill, Wilmington, Delaware.

♦ **FIGURE 8-12.**　Descriptive copy creates customer interest in the menu from Cascades restaurant at the Opryland Hotel, Nashville, Tennessee. Reprinted with permission of Opryland Hotel, Nashville, Tennessee.

♦ **FIGURE 8-13.**　The descriptive copy for this menu from Wales, Great Britain is often amusing and to the point.

pasta. What might have simply read "Mussels with peppers, butter, bacon and pasta" now reads:

Mussels Casino
 Steamed Prince Edward Island mussels with bell peppers, casino butter, apple smoked bacon and fusilli pasta. . .
Grilled Chicken Salad
 Wilted greens, roasted corn and garlic-scented mushrooms tossed with a cherry pepper remoulade. . .
Grilled Salmon Salad
 Over mixed greens with pickled fennel and cucumber, vine-ripened tomato and fennel-mustard vinaigrette. . .

♦ **FIGURE 8-14.** Figures 8-14 through 8-18 use food associated words to enhance the menu item description.

Peppered Roast Veal Loin
 with pine nuts, sun-dried tomatoes and Iowa blue cheese sauce Artichoke, mushroom and smoked Cheddar terrine with black olive puree

---

♦ **FIGURE 8-15.**

The description for seafood au gratin in Figure 8-19 is simple, but very effective:

Matt's Seafood Scrod Au Gratin
 Over 1/2 lb of fresh haddock or cod baked in a rich dill sauce and mozzarella cheese . . . priced daily

---

♦ **FIGURE 8-16.**

This description for clams uses the phrase "deep fried, full-bellied, lightly breaded" to create the picture of hot, crisply fried, plump, juicy clams:

> Fried Clam Plate
>   Deep fried, full-bellied clams, lightly breaded, with our french fries & cole slaw. . .

♦ **FIGURE 8-17.**

Goulash is the subject of this menu item description. Reading this, a customer could easily smell the heavy aroma of beef stock and vegetables that have been slow cooked into a rich and flavorful soup:

> HUNGARIAN RHAPSODY
> A spectacular Hungarian Goulash Soup. A thick, rich beef stock simmering with red-ripe tomatoes, fresh garden vegetables, chunks of tender beef, imported herbs and spices and laced with dry sherry, then topped with a dollop of sour cream. . .

♦ **FIGURE 8-18.**

TRUTH IN MENU: THE LEGAL ASPECTS OF THE PRINTED WORD

By law, a menu is classified as advertising. Whichever type of menu copy is used—merchandising, accent, or descriptive—the written words on a menu are intended to increase sales by presenting the restaurant product in the best possible light. In an effort to impress customers with the quality of the ingredients and the style of preparation, menu designers and planners are often guilty of claiming that an item is something that it is not. This practice can lead to two situations that are detrimental to the foodservice operation: violation of truth-in-advertising law and encouraging false expectations in the customer.

While misrepresentation can occur in merchandising and accent copy—if, for example, the private room offered by the restaurant is described as something that it is not—descriptive copy is the area that must be monitored most carefully. Legally, each item advertised on the menu must be what the menu claims it is. Violations of truth in menu occur most frequently in the following ways in descriptive copy.

Grades of Meat, Produce and Dairy Products

All meats are inspected by the U.S. Department of Agriculture and graded according to quality standards such as prime and choice. To serve a different or lesser USDA grade than the grade printed on the menu is considered misrepresentation, or fraud. Similarly, fruits, vegetables, and dairy products are also USDA graded with terms such as Grade A, Grade AA, Fancy, and Extra Fancy. If the menu planner decides to use these terms in descriptive copy, the foods must meet those standards.

Size and Weight of Foods

Meat entrees often specify the weight and size of a cut of meat. Traditionally it has been understood that size and weight specifications are determined before the meat is cooked. Recently, however, some fast-food chains have had to post the after-cooking weight and size of their products because of customer demand. Inferior-quality blends of ground beef have proven to be the principal offenders in substantial discrepancies between raw and cooked product.

Because of fluctuating prices, it is suggested to avoid specifying the weight of an item on the menu unless absolutely necessary. If the menu does specify weight, the restaurant may have to lower the quality of meat used in order to meet the weight posted on the menu if cost of product increases sharply. If the weight is omitted from the menu, however, the restaurant may decrease the weight but continue to use the same quality product.

Some foods are also graded, according to size, such as shrimp and eggs. As with other official regulations, if the menu claims that the item is a specific regulation size, the item must meet official standards. Extra large shrimp cannot be substituted for jumbo shrimp.

Type of Product and Degree of Freshness

Truth-in-advertising laws are quite clear in this regard. If the menu says that an item is fresh, it must be fresh, not canned, frozen or "fresh-frozen." If a fresh fruit salad were offered as an appetizer, it would be considered misrepresentation if the fruits were canned. Similarly, if the accent copy lists an item as sirloin tips, an inferior cut of meat cannot be substituted.

Geographical Origin of Food

Whether an item is imported or domestic can have a great impact on the sale of the product. However, it is illegal to make false claims about the geographical origin of a product. Imported Swiss cheese does not come from Wisconsin. Italian provolone must be from Italy. Different types of seafood can fall into this category: Alaskan king crab, for instance, is different from blue crab.

References to regional styles of preparation and origin must be worded carefully. For example, Virginia ham must be from Virginia, but Virginia-style ham refers to a style of preparation. New England clam chowder, however, is understood to refer to a style of preparation, just as Manhattan clam chowder does not necessarily mean that the soup was made in New York City.

Style of Preparation

Customers often base their menu decisions on the way that an item is prepared. Diet-conscious customers will be annoyed if the item they ordered is served fried instead of broiled, as the menu claimed. In addition, marketing efforts to enhance a product through the use of such terms as homemade can cause problems. The term homemade implies that the item has been made in a private home kitchen; most health departments have licensing restrictions against the production of restaurant food in private, uninspected home kitchens. Home-style is a much better adjective to suggest the attributes of a homemade dish.

Dietary and Nutritional Claims

The general public has become more aware of artificial additives as well as the nutritional levels of many foods. Menu designers and planners should be especially carefully when claiming that menu items have certain nutritional benefits or have limited amounts of additives. The exact amount of each nutrient or ingredient must be available to customers on demand. Increasingly, menus are featuring "heart healthy" or "green" or "light" items. Management must be aware that there is a legal liability on the part of the restaurant if these items are not prepared as advertised on the menu.

Customer Expectations

Anyone who has ever ordered a tempting-sounding veal parmigiana and been served a veal patty rather than a veal cutlet will understand the need for menu planners to be scrupulous when writing descriptive copy. Although the writer of the copy may be within the legal limits when making a description sound appealing, it is not to the restaurant's advantage to encourage false expectations on the part of the customer. The following menu copy provides a good example: "A steaming crock of country-style French onion soup, heavy with onions and crusty bread and topped with melted cheese." If the customer is presented with a soup bowl of tepid, pureed onion soup sprinkled with a bit of grated cheese, they will be justifiably disappointed. The customer was promised a steaming-hot crockery dish, authentic Provencal-style soup with generous amounts of onions, and the classic presentation of crusty bread soaked in soup and topped with melted cheese.

The menu planner must be sure that everyone who will be involved in introducing the item is in agreement about the preparation and presentation. Otherwise, the item may not live up to its description. In addition, the use of superlatives such as magnificent or to perfection also encourages high expectations. No one likes to be disappointed. When writing descriptive copy, be aware of what the restaurant staff is capable of producing.

Efforts to control truth in menu are being made by the National Restaurant Association and its local chapters. Ultimately, their efforts should help to produce menus that more accurately represent menu items to customers. The general public will also become assured of the foodservice industry's professionalism and concern for customer satisfaction.

Using Foreign Languages in Menu Copy

Americans have always reacted favorably toward foods, styles, and cuisines of other countries. French food dishes, in particular, have had great appeal, primarily because French cuisine is considered the classical foundation of gourmet food preparation. The impact of international cuisines on the American foodservice industry can be measured in the wide range of cuisine restaurants present in most communities and in the number of international language phrases and terms relating to foods and preparation methods used on menus. A review of the menus featured throughout this book will reveal French, Italian, German, Gaelic, Japanese, Chinese, Polynesian, Australian, Spanish, Mexican, Swedish, Hungarian, Russian, Indian, and Greek words and phrases.

Using a foreign language to sell a menu is a good marketing tool. However, the menu planner must be sure to provide an explanation of exactly what the dish is if not regionally or nationally well known. If no explanation is provided and the customer has to ask to have the menu interpreted, the customer may become annoyed or embarrassed and pass over the dish, which may not help restaurant sales and may inhibit the quality of their experience. An American customer faced with the French table d'hote menu in Figure 8-19 might not recognize the menu items. The menu in Figure 8-20, which uses German for marketing purposes and explains the items in simple, interesting copy, is far more appropriate for the American customer.

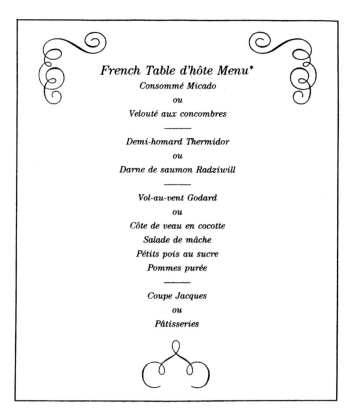

French Table d'hôte Menu*
Consommé Micado
ou
Velouté aux concombres

Demi-homard Thermidor
ou
Darne de saumon Radziwill

Vol-au-vent Godard
ou
Côte de veau en cocotte
Salade de mâche
Pétits pois au sucre
Pommes purée

Coupe Jacques
ou
Pâtisseries

◆ **FIGURE 8-19.** A table d'hôte menu in French.

Fischgerichte (fish)

frisch aus dem Netz und von der Angel...
the Old Brauhaus proudly offers a variety of the freshest seafood and fresh water fish available
...ask your server about todays catch!

Krebskroketten
fried deviled crab made from a family recipe using lump backfin and alaskan snow crabs

Jakobsmuscheln überbacken
sauteéd fresh sea scallops in a creamy riesling wine sauce over egg noodles with grüyere cheese au gratin

Lachsscheiben gegrillt
grilled north pacific salmon steak topped with a dill hollandaise

Tiefsee Garnelen
succulent jumbo gulf shrimp broiled in herb butter or lightly breaded and deep fried

Meeresfrüchte Varieté
an assortment of fresh seafood broiled in lemon butter or lightly breaded and fried

"Matrosen an Land"
a 6-8 oz. broiled new zealand lobster tail, two fried jumbo gulf shrimp and a petite filet mignon topped with herb butter...
...the best of both worlds!

Geflügel (poultry)

Hühnerschnitzel mit Pinienkerne
lightly breaded breast of chicken sauteéd with raisins and pine nuts

Gefüllte Hühnerbrust "Cordon Bleu"
boneless breast of chicken stuffed with westphalian ham and vermont cheddar cheese

vom Schwein (pork)

Schweinebraten
a favorite german dish of roast pork loin with mashed potatoes and sauerkraut

Zigeuner Schnitzel
sauteéd pork cutlet...topped with julienned onions, sweet bell peppers and mushrooms, in a spicy hungarian paprika sauce

vom Kalb (veal)

the Old Brauhaus uses the finest milk-fed veal available...hand cut by our chef to insure its quality and freshness.

Wiener Schnitzel
the ever popular vienniese dish sauteéd in sweet butter

Kalbsschnitzel mit Käse und Schinken
sauteéd veal topped with smoked ham and muenster cheese

Geschnetzeltes vom Kalb
slivers of veal sauteéd with black forest mushrooms in a creamy trollinger wine sauce

Kalbsleber "Linderhof"
sauteéd tender calves liver in a düsseldorf mustard horseradish sauce

♦ **FIGURE 8-20.** German is used in this menu for accent copy and then translated for descriptive copy.

When using a foreign language in menu copy, the general rules of grammar for that language should be followed. In French, capital letters are used only for the first letter of the word in a sentence and for proper nouns—names of people, places, and styles of preparation. In the phrase Demi-Thermidor, the first letter of the phrase is capitalized and the "t" in thermidor is capitalized because it is a method of

ANTIPASTI

**Salmone E Branzino Marinati Al Pepe Rosa** Salmon and seabass marinated with pink pepper corn in olive oil dressing	P 240
**Prosciutto E Melone** Thin slices of prosciutto ham with melon	P 315
**Scallopa Di Fegato Grasso All'Aceto** _**Balsamico E Indivie Dorate**_ Pan-fried goose liver served on a bed of golden witlof salad and balsamic vinegar	P 450
**Carpaccio Di Manzo Alle Lattughe** Beef carpaccio with creamy lettuce sauce	P 310
**Scampi Ed Asparagi Con Crema D'Olio** _**D'Oliva Piccante**_ Scampi and asparagus with spicy olive oil cream	P 450
**Cernia Al Piatto** Thin slices of red snapper with fresh tomato basil, thyme, olive oil and lemon zest	P 230
**Guazzetto Di Triglie Allo Sherry** Pan-fried red mullet in sherry wine broth	P 290
**Pomodoro E Mozzarella** Salad of mozzarella cheese with slices of tomato and basil	P 260
**Insalata Mista Di Stagione** Fresh seasonal salad	P 145

ZUPPE

**Minestrone Di Verdura** Fresh vegetable soup	P 125
**Zuppa Di Cipolle Alla Pavese** Italian onion soup with egg yolk swirls	P 140
**Crema Di Patate E Pollo Al Profumo** _**Di Rosmarino**_ Potato and chicken cream parfumed with rosemary	P 150

♦ **FIGURE 8-21.** Italian is used for the accent copy in this menu.

preparation. In German, the umlaut accent mark used over certain vowels is as important to the menu's authenticity as the French accent marks.

The Italian menu shown in Figure 8-21 features the appetizer and soup courses and uses a number of Italian words commonly found on English language menus. The Hebrew menu in Figure 8-22 is an excellent example of an international menu

◆ **FIGURE 8-22.** Hebrew and English are featured in this menu from the Moriah Plaza Hotel Dead Sea. Reprinted with permission of Radisson Moriah Plaza Hotel Dead Sea.

printed in two languages. As each language has its own column on the menu, customers do not get confused while making menu selections. Norwegian is the featured language on the menu in Figure 8-23 from the Radisson SAS Park Hotel in Oslo, Norway. The English translation is well written and easily distinguished from the Norwegian version. Italian along with English food words are blended into the Norwegian menu item descriptions.

The Japanese menu in Figure 8-24 is an excellent example of the mixed use of Japanese text, the English text version of the Japanese, and the English translation for all of the menu items. The Japanese text provides the accent copy for this menu.

FORNEBU
PARK
RESTAURANT

STARTERS

Gravet laks med sennepssaus, servert med dillstuede poteter. 95,-
Dill marinated salmon with mustard sauce & dill stewed potatoes.

Kylling grønnsaksterrine med tomatvinagrette. 64,-
Chicken vegetable terrine with tomato vinagrette.

"Toast Midnatt Sol," med limfjordskaviar, løk & eggeplomme. 84,-
Toast midnight sun, with caviar from Limfjord, onions & egg yolk.

SALADS & SOUPS

Cesar salat med lunt kyllingbryst & parmesan ost. 86,-
Caeser salad with warm chicken breast & parmesan cheese.

Issalat garnert med bacon. 42,-
Ice berg salad garnished with bacon.

Skalldyrsuppe med aiolisaus. 68,-
Cream of shrimp soup with aioli sauce.

Minestronesuppe med pesto. 48,-
Minestrone soup with pesto.

PASTA & LIGHTMEALS

Fettucini med røkt laks, reker & hvitvinsaus. 98,-
Fettucini with smoked salmon, shrimps and white wine sauce.

Pasta med friske grønnsaker på en lett tomatsaus med basilikum. 85,-
Vegetable pasta with a light tomato sauce & fresh basil.

Club sandwich med bacon, kylling, egg & potetchips. 92,-
Our famous club sandwich with bacon, chicken, egg & crisps.

150gr. hamburger med cole slaw & pommes frites. 76,-
150 gr. hamburger with cole slaw & french fries.

PRØV VÅR NYE FONDUE BUFFET HVER FREDAG &
LØRDAG KR .175,-
TRY OUR NEW FONDUE BUFFET EVERY FRIDAY &
SATURDAY. KR 175,-

RADISSON SAS PARK HOTEL OSLO, Fornebuparken, N-1324 Lysaker, Norway
Telephone; +47 67 12 02 20 Fax: +47 67 12 00 11

♦ **FIGURE 8-23.** English is the primary language on this menu with Norwegian providing the accent copy. Reprinted with permission of Radisson SAS Park Hotel Oslo, Norway.

YAKIMONO
焼物
Grilled dishes

KURUMAEBI SHIOYAKI 車海老塩焼	Grilled prawns	₱ 250
SAWARA SHIOYAKI 鰆 塩 焼	Grilled fish in season	180
SAIKYO YAKI 季節魚西京焼	Grilled fish marinated with soya bean paste	210
IKA TERIYAKI いか照り焼	Grilled cuttlefish with sweet soya sauce	210
SHAKE SHIOYAKI 鮭塩焼	Grilled salmon	400
SHAKE BUTTERYAKI 鮭バター焼	Pan-sauteed salmon with butter	400
UNAGI KABAYAKI 鰻蒲焼	Grilled eel	550
WAKADORI KUWAYAKI 若鶏桑焼	Grilled spring chicken with garlic-flavored soy sauce	210
WAKADORI TERIYAKI 若鶏照り焼	Grilled chicken with sweet soy sauce	210
ISEEBI ONIGARAYAKI 伊勢海老鬼殻焼	Grilled lobster "ONIGARA" style	300/100 grm.
GINDARA TERIYAKI 銀鱈照焼	Grilled codfish with teriyaki sauce	300

SASHIMI/TSUKURI
刺身・お造り
Sliced raw fish

BENKAY SASHIMI MORIKOMI 弁慶刺身盛り込み		
(MATSU) for 5 松（５人前）	Special assorted sashimi (for 5)(local/imported)	₱ 1,300
(TAKE) for 3 竹（３人前）	Special assorted sashimi (for 3)(local)	700
SASHIMI GOTEN MORI 刺身五点盛り	Five kinds of sliced raw fish (local)	300
SASHIMI HATTEN MORI 刺身八点盛り	Eight kinds of sliced raw fish (local)	420
MAGURO 鮪	Tuna	220
SHIMESABA しめさば	Marinated mackerel	295
LAPU-LAPU USUZUKURI ラプラプ薄造り	Thinly-sliced lapu-lapu with ponzu sauce	220
UNI うに	Sea urchin	220
IKURA いくら	Salmon roe	450
IKA いか	Cuttlefish	220
SHAKE 鮭	Salmon	450
HAMACHI はまち	Yellow tail fish	450

◆ **FIGURE 8-24.**   This menu is an excellent example of the use of Japanese and English to create a multi-language use menu. Reprinted with permission of Dusit Hotel Nikko, Manila, Philippines.

OVERVIEW

As in the chapter on menu design, a number of topics and issues related to the use of copy have been discussed. When all of the copy for the menu design has been determined and laid out on a graphic format, ask the following questions to help you evaluate just how well your customers will be able to understand and react to the merchandising, accent, and descriptive copy on your menu.

◆ Does the written copy present the menu in an interesting and/or entertaining manner?

◆ Have you adequately described those menu items which need to be explained to the customer?

◆ PART IV ◆

Marketing For Profit

CHAPTER 9 ◆

Restaurant Concept and Cuisine Trends

Marketing is selling. Salad bars and welcome buttons are obvious applications of marketing programs. More complex approaches include analyzing unit sales from sales mix information to develop promotions directed at customer preference trends.

Marketing is developing customer sales by reacting to customer demands and understanding how trends in the demographic and social patterns of a community will impact an individual foodservice operation. Marketing requires the knowledge of how much the customers have to spend and what their dollars will be spent on.

Marketing is the study of what can be done with an operation to expand turnover rates, increase sales, and finally to increase profit percentages. Marketing can be as simple as creating a Wednesday night theme buffet or as complex as redirecting the entire focus of a foodservice organization. In the following section, current marketing trends in the foodservice industry, new item development, and the initial consideration of advertising for commercial operations are discussed.

Management can use marketing as a key tool for sales development by recognizing and analyzing current customer buying trends, volume sales patterns, and overall cost percentages, as they affect the profit margin.

Competition for the American restaurant customer in the 1990s was fast and furious. The National Restaurant Association estimates that consumers will spend $336 billion dollars in food consumed outside of the home in 1998. Of every dollar spent on food in the United States, 44 cents is spent in restaurants. (Source National Restaurant Association.) Figure 9-1 details the breakdown of consumer spending in specific segments of the restaurant marketplace.

The segment that has seen the greatest growth is casual family restaurants to answer the changing needs of the American family. Chief among the established national chains featuring casual family-style are Bennigan's, T.G.I. Friday's, The Ground Round, Olive Garden, Red Lobster, Chili's, and Pizza Hut. Restaurant chains that have made significant inroads into this market in the 1990s include Applebee's, Macaroni's, Lonestar, and Longhorn. In addition, many independent operators are developing restaurants along the same casual theme and offering similar menu items. The menus from a number of the national and independent casual-style restaurants are featured throughout this book.

A recent article in _Bon Appetit_ magazine focused on the casual-style restaurant and identified a bistro, a trattoria, a steakhouse, and a cafe as the feature "Casual-Style" restaurants for this issue. William Garry, _Bon Appetit_ Editor-in-Chief, said of the casual-style restaurant, its themes, and cuisines: "What is the nature of this new Americanization? It is eclectic in its food preparation, friendly in its service, democratic in its welcome and casual in its overall personality. It has embraced all strains of other types of eating establishments, from the diner to the ethnic place to the country inn. It has the sophistication of its European prototype, married to the relaxed feeling of being at home."

The fast-food industry has developed a "quick food" segment in answer to the growing need for "home replacement food." Boston Market and Kenny Rogers are two national chains that have positioned themselves to answer customer needs in this area. Both restaurants offer fully prepared poultry and beef meals available in single or multiple servings. Restaurants for both companies supply comfortable dining areas for customers to eat in as well as microwaveable packaging for take-out. The menus for both are designed to provide "home-style" foods such as roasted

Consumer Spending By Food Service Segment

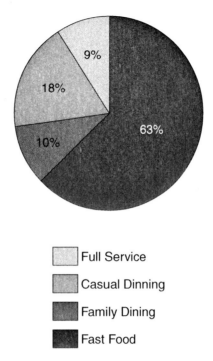

Full Service

Casual Dinning

Family Dining

Fast Food

♦ **FIGURE 9-1.** Consumer spending patterns in the restaurant industry.

chicken, baked ham, meatloaf, roast beef, and turkey, along with the traditional side items—mashed potatoes, vegetables, stuffing, salads, and accompaniments. Dessert is available, usually seasonal pie. Heartly sandwiches are also available, incorporating the same food items. Observing customers in these operations can be a revealing look at how American families are adapting to changing lifestyles and work patterns. Recently a multigenerational family was observed at a Boston Market restaurant sitting down to a "family dinner" with roast chicken and ham as the main course. Eight people were seated around the tables, the grandfather at one end and a baby in a high chair at the other. On either side were seated the balance of parents, siblings, and grandparents.

The pizza marketplace has expanded into casual restaurants centered on pizza as the focus of a menu offering a wide range of items. Bagel-oriented restaurants have also seen tremendous growth nationally. Figure 9-2 is the menu from Einstein Brothers Bagels. Restaurants open in the early morning and operate into the evening, serving bagels as breakfast, sandwiches, and pizza along with soups, salads, desserts, and gourmet coffees for both take-out and in-house consumption. In addition, operations also offer the catering menu seen in Figure 9-3.

Adding new items to the menu is done primarily to increase the average check, attract new customers, and develop the present customer market. The reasons for

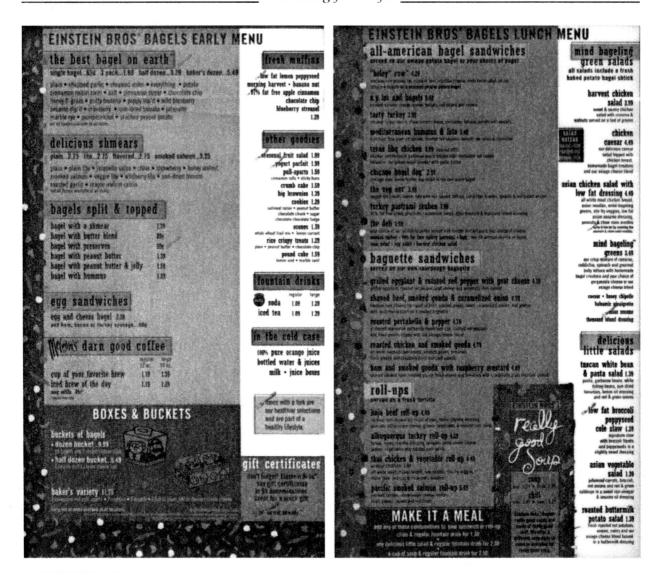

◆ **FIGURE 9-2.** The trend in bagel-oriented quick-serve restaurants is reflected in this menu from the national chain, Einstein Brothers Bagels. Reprinted with permission. © 1998 Einstein Noah Bagel Corp.

offering a new item should be established before product and price decisions are made. If products will increase the average check, the pricing should be kept high. If the product is intended to attract new customers, the price should be kept down and increased profits realized through volume sales or low-cost items. If the product is supposed to develop the existing market, the price should be based on what customers will perceive as fair.

Whatever the new product, it should be compatible with the existing menu. Adding an elaborate selection of appetizers to a soup, salad, and sandwich menu,

Gift Baskets
Assorted Baker's Dozen Bagels
2 Tubs of the World's Best Cream Cheese
(1 Tub of Plain and 1 Tub of Lite or Flavored Cream Cheese)
1 bag of Bagel Chips
1 bag of Einstein Bros. Signature Coffee **$29.**⁹⁵

Bagel Bucket
Assorted Baker's Dozen Bagels
2 Tubs of the World's Best Cream Cheese
(you choose the flavor) **$8.**⁹⁹

Bagel Cream Cheese Tray
1 Dozen Bagels (up to 3 different varieties)
with your choice of four Cream Cheese flavors
cut in 1/2's or 1/4's **$19.**⁹⁵

Combination Platter
Bagel Deli Sandwiches and Bagels with schmear
6 Bagel Sandwiches
Choose up to three varieties: Turkey, Turkey Pastrami, Corned Beef,
Ham, Tuna, Chicken Salad or Whitefish Salad.
Cheese: American, Swiss, Cheddar or Provolone.
Lettuce, Tomato, Mayonnaise or Mustard
6 Bagels spread with your choice of Cream Cheese
Cut in 1/2's or 1/4's **$32.**⁹⁵

Deli Sandwich Platter
1 Dozen Bagels (up to 3 different varieties)
Choose up to four varieties: Turkey, Turkey Pastrami, Corned Beef,
Ham, Tuna, Chicken Salad or Whitefish Salad.
Cheese: American, Swiss, Cheddar or Provolone.
Lettuce, Tomato, Mayonnaise or Mustard
Cut in 1/2's or 1/4's **$44.**⁹⁵

Side Salads:
Cole Slaw, Potato or Pasta Salad (32oz) .... **$4.**⁹⁵ to **$5.**⁹⁵

Boxed Lunch
Choose any Deli Sandwich
Choice of Side Salad or Potato Chips
Cookie: Oatmeal, Peanut Butter, Chocolate Chip or Sugar   **$5.**²⁵

Choice of Coke, Diet Coke, Sprite, Iced Tea, Bottled Water **$6.**²⁵

♦ **FIGURE 9-3.**   By expanding the regular menu, Einstein Brothers Bagels offers catering services to homes and businesses. Reprinted with permission. © 1998 Einstein Brothers Bagel Corp.

221

for example, would not be advisable. Similarly, adding Mexican food to a New England cuisine menu would not necessarily be successful. Adding Shaker foods to the New England cuisine menu would, on the other hand, probably result in increased sales. Regionality has a great deal to do with how customers accept cuisine-oriented menu items. Any new menu item will cost management time and money. The need for a new menu product should be clearly identified before is it developed.

CUISINE TRENDS

From nouvelle to southwestern, from pasta to Cajun, from fusion to Asian, from Mediterranean to Caribbean, a whole wave of cuisine and dining trends has swept through the restaurant industry, often leaving behind broken bank accounts and empty store fronts. But many of the new generation of American restauranteurs and chefs have carved out a place for themselves and their customers, creating unheard of combinations and brilliant restaurant decors. Along the way they have managed to revolutionize us all as their trends and cuisines spill over onto even the most traditional menus.

Cuisine trends reflected in many restaurant menus today, in addition to the traditional European cuisines such as French, Italian and German, include:

◆ Southwestern
◆ Mediterranean
◆ Asian
◆ Cajun
◆ Mexican
◆ Caribbean
◆ Indian
◆ Lean, Healthy, Lite, or Heartful

The menu from the Iron Hill Brewery and Restaurant in Newark, Delaware featured in Figure 9-4 reflects the new look in American casual-style restaurants. The target market for this restaurant is fairly sophisticated and receptive to current food trends. The restaurant is located in a university town, partway between Philadelphia, Pennsylvania, and Baltimore, Maryland, and draws on a customer base that includes professionals, university students and staff, workers from a wide range of neighboring industries, along with visitors and local residents. Based on a microbrewery theme, the menu blends together Asian, Italian, Mediteranean, Cajun, Caribbean, and traditional American cuisines. The list of featured pizza is a summary of the menu's cuisine versatility.

American cuisine is represented in major American cities around the country by chefs who concentrate on using fresh, natural/organic local food products with an emphasis on presentation and cuisine. Often the cuisine is indigenous to the local region. Sometimes chefs will combine regional ideas or transplant themselves to other areas of the country, carrying their regional cuisine with them. The results

DINNER ENTREES

Spicy Vegetable and Penne Stirfry 10.50
Oriental vegetables in a spicy Cantonese-style sauce

Rock Shrimp and Andouille Sausage Etouffee 13.50
classic sauce made with Pig Iron Porter, over rice

Babyback Ribs 10.95 / 15.95
green chili cornbread and cole slaw

Rosemary Chicken with Tomato Linguine 13.50
oyster, shiitake and crimini mushrooms in sherry
wine sauce with lemon and fresh rosemary

Atlantic Salmon Filet 14.50
chef's daily preparation

Iron Hill Meatloaf 12.50
grilled, with garlic mashed potatoes, smoked
mushroom duxelle and demi-glace

Baked Salmon Cakes 14.50
quinoa and roasted corn pilaf with green onion
and dill remoulade

Wood Oven Roasted Half Chicken 12.50
marinated in lemon and fresh herbs,
with garlic mashed potatoes

Jamaican Jerked Pork Chops 14.95
pineapple-bell pepper salsa and sweet potato fries

Charbroiled New York Sirloin 16.95
12 ounce cut of angus beef with sauteed button
mushrooms and baked potato

Shrimp Ravioli 13.00
artichokes and leeks in a sage-infused tomato broth

SANDWICHES

served with vegetable slaw and french fries

Smoked Turkey Breast 6.95
served on honey wheat bread with havarti cheese, apricot
whole grain mustard and alfalfa sprouts

Grilled Chicken Breast 6.50
roasted red peppers, marinated tomato, red onion and
basil aioli on focaccia

Carolina Pork Bar-B-Q 5.75
in pepper vinegar sauce with cole slaw on a snowflake roll

Pub Style Roast Beef 7.50
sliced thin with sharp cheddar, lettuce, tomatoes and
horseradish-chive dressing on onion loaf

Brewski Burger 6.50
half pound Angus ground beef on a garlic pretzel roll with
mushrooms, bacon and choice of provolone, swiss or cheddar

WOOD OVEN PIZZA

Southwestern 9.50
grilled chicken, black beans,
salsa, jalapenos and cilantro,
with cheddar and jack cheeses

Mediterranean 8.25
feta, imported olives, plum
tomato, garlic and basil

Lejon 10.50
rock shrimp, horseradish, bacon,
scallions and mozzarella

Grilled Vegetable 7.95
zucchini, eggplant, yellow squash,
red onion, tomato and mozzarella

Brew House 8.50
grilled sausages, bell peppers,
onions and mozzarella

Garcia 8.95
wild mushroom, plum tomato
and gorgonzola

Traditional 7.50
tomato sauce, fresh basil
and mozzarella

Additional Toppings 1.00
pepperoni, sausage or mushroom

BEVERAGES

Iron Hill Rootbeer 1.75
Coke, Diet Coke, Sprite, Ginger 1.50
Perrier, Evian 1.75
Coffee, decaf and hot tea 1.50
Iced tea 1.50

*Cigarette, cigar and pipe smoking permitted in
the bar and billiards room only. A 17% gratuity
may be added to parties of six or more.*

DINNER MENU

*DINNER IS SERVED FROM 5:00 PM UNTIL 9:00 PM SUNDAY THROUGH THURSDAY
5:00 PM UNTIL 10:00 PM FRIDAY AND SATURDAY*

SOUPS AND CHILI

Black Bean Soup
topped with chopped onions and jalapeno
2.95 / 3.95

Thai Fish and Vegetable Soup
fresh daily selection of fish and a variety
of vegetables,with lemongrass, ginger,
chili peppers and cilantro
3.50 / 4.50

Chicken and Sausage Gumbo
hearty cajun stew with andouille, okra,
rice and authentic seasonings
3.95 / 4.95

Numero Uno Chili
spicy recipe with ground beef and kidney
beans, served with monterey jack,
jalapeno and corn tortillas
3.50 / 4.50

SALADS

served with homemade breadsticks

Tossed Green Salad 3.50
mixed greens, grated carrots, cucumber,
red onion and plum tomatoes served
with fresh herb vinaigrette

Mesclun Salad 4.95
fresh spring mix, red onion, plum
tomatoes, pancetta, herbed chevre
croustades, with fresh herb vinaigrette

Caesar Salad 3.95 / 5.95
crisp romaine and freshly made dressing
with croutons and parmesan cheese
julienne of grilled chicken breast 8.95
marinated portobella mushrooms 7.95

Santa Fe Salad 8.95
romaine, grilled marinated chicken,
black beans, corn, tomatoes, cilantro,
julienne corn tortillas, grated cheddar
and monterey jack cheeses served with
a creamy salsa or spicy peanut dressing

Warm Sirloin Salad 9.50
marinated in Ore House Amber
with romaine, radicchio, plum tomato,
bell peppers, red onion, olives, fresh
mozzarella and balsamic vinaigrette

APPETIZERS AND SHARED PLATES

Tex-Mex Egg Rolls 6.50
grilled chicken, black beans, corn, red onion,
peppers, with diced tomato and avocado cream

Stuffed Kennett Square Portobella 7.50
roasted peppers and stewed lentils with smoked ham
and leeks, baked with fontina cheese

Grilled Sausage Sampler 6.95
andouille, knockwurst and garlic kielbasa with white bean
ragout served with horseradish, dijon and grain mustards

Black and Tan Hummus 5.50
spicy black bean puree paired with garbanzo bean puree
served with grilled soft pita and vegetable stix

Jack Straw Onions 3.95
thinly sliced onions tossed in seasoned flour,
deep fried golden brown

Chicken Wings 6.50
choice of smoky pepper or classic buffalo sauce
with blue cheese dressing and vegetable stix

Hot Artichoke and Cheese Dip 5.75
baked and served with toasted beer bread

Bruschetta 4.95
chef's daily selection

Guacamole and Tortilla Chips 4.50
fresh avocado, tomato, cilantro and jalapenos

Fresh Salsa and Tortilla Chips 3.50
made daily with tomatoes, onions, jalapenos and cilantro

House Nachos 8.50
piled high with diced tomatoes, black beans, green
onions, monterey jack and cheddar cheeses served
with salsa, sour cream and jalapenos

◆ FIGURE 9-4. The menu from the Iron Hill Brewery and Restaurant, Newark, Delaware, blends together a wide range of cuisines. Reprinted with permission of Iron Hill Brewery and Restaurant, Newark, Delaware.

are a blend of national cooking techniques and products represented internationally as American cuisine. Blended in are traditional American dishes such as chicken pot pie, meatloaf, roast turkey and stuffing, fried chicken, prime rib of beef, baked ham, roast loin of pork, and salmon steak. Perhaps for the first time, American cuisine is being taken seriously around the world as an identifiable group of cooking techniques, food items, and tastes.

Southwestern Cuisine

What began slowly as a food trend has become an established part of American restaurant menus throughout the United States. Fajitas, nachos, and taco salad are just three of the many southwestern cuisine menu items found on many menus today. The American restaurant customer responded so favorably to both foodservice and retail outlets offering southwestern-style foods that menus soon started offering salads, soups, entrees, appetizers, and sandwich items with Mexican names and ingredients. One of the major restaurant chains to develop this theme cuisine is Chili's, whose menu is featured in Figure 7-10b. At the same time, full-service restaurants from New York to San Francisco attracted chefs who experimented with the fine dining elements of southwestern cuisine. The result is a broad spectrum of restaurants offering every level of this cuisine, which reaches into the rich history of Mexican cuisine. These developments established southwestern cuisine as a permanent part of the American culinary portfolio, along with New England, Cajun, mid-Atlantic, Californian, and midwestern styles.

Asian Cuisine

Asian cuisine goes far beyond the Chinese take-out and the Japanese steakhouse to encompass a combined group of cooking techniques, preparation styles, and ingredients from the Asia Pacific Rim. From Thailand to Tokyo, from Beijing to Singapore, the multitude of eclectic spices and foods are at the mercy of today's American chefs as they search for new tastes, presentations, and techniques to incorporate into their menus.

Home Replacement Food

Once known as "take-out to eat-in," this market has been formalized to cover the range of food items outside of the fast-food market, offered fully prepared for home consumption. Establishments from local delicatessens and charcuteries to the meat section in local supermarkets make available ingredients that the consumer would normally have purchased as fully prepared food items. The growing market for this segment continues to be fueled by the two-income family and the increasingly diminishing amount of time available for at-home food preparation. Given the choice between an hour of "quality" time with their families or shopping for and preparing a meal, customers are increasingly making the decision to turn over the responsibility for food preparation to a local service provider. Associated with this trend is an at-home restaurant food delivery system, known in many areas of the country as 'Takeout Taxi' (Figure 9-5). Local restaurants sign onto the service, identify good menu items for home delivery, and let the service do the delivery system. The benefit for the restaurants is revenue that would ordinarily been lost to another provider. The customer gains with the convenience of having their favorite restaurant food "at home."

Lean/Healthy/Heartful

As discussed previously, an essential element in successful marketing is knowing how customers want to spend their dollars. Healthy eating has become a way of life with many Americans, extending beyond the home to their local restaurants. Frequent travelers have also increased the demands on hotel restaurants to provide similar healthy menu items. The menu in Figure 8-9 from the Sheraton Hotel at Disney

♦ **FIGURE 9-5.** Home delivery services such as the national franchise, Takeout Taxi, expand the consumer market within a local area for many restaurants. Reprinted with permission of Takeout Taxi.

World provides nutritional notes for each menu item. Customer concerns in this area include a demand for:

♦ Fresh food products versus preprepared items using preservatives and other additives
♦ Menu items low in calories and fat content, prepared in ways that reduce the use of oil and salt
♦ The use of light cooking methods such as steaming, poaching, and grilling

These three concerns, viewed together, indicate a concentrated demand for fresh, low-calorie, natural food items and represent a growing trend that menu planners should be aware of.

Business Breakfast

Breakfast has gained an important place in marketing strategies. Weekend and holiday brunch has become an increasingly popular meal. The business community has found that a successful way of doing business is around the breakfast table. The early meeting leaves more business hours free, while the lighter, nonalcoholic menu is more conductive to productivity than long lunches or after-hours cocktail meetings. The business community has been affected as much by economic pressures to cut down on entertainment expenses as by current health and fitness trends. Breakfast meetings are less expensive than lunches and dinners and allow more time in the day for exercise and other health-related pursuits. Furthermore, marketing the breakfast meeting is becoming a new way to sell restaurant and meeting-room space, which can help foodservice operations offset lower average luncheon and dinner checks.

Recognizing Trends

In recognizing different trends, restaurants can develop menus to respond to the needs of their customers. New trends are always in the making. By being aware of what is happening with consumer spending patterns on the local and national levels, operators can spot trends and new marketing concepts for the foodservice industry. The National Restaurant Association publishes several reports to keep its members up to date on current trends. In addition, an annual report provides an overview of the restaurant industry nationally. A number of industry-related and consumer magazines also provide invaluable information about what is happening in restaurants around the United States. *Restaurant and Institutions* is just one of the trade publications available by subscription to foodservice operators. *Food & Wine, Bon Appetit, Eating Well, Gourmet, Saveur, Culinary Trends, Coffee Journal,* and *Wine Spectator* are just some of the consumer magazines which provide professional-level information.

CHAPTER 10

♦

Specialty
Menus

◆ **FIGURE 10-1a, b.** The seasonal summer menu from The Ground Round restaurants offers a selection of sandwiches and grilled items. Reprinted with permission of Ground Round Restaurants.

SEASONAL MENU

One of the most interesting trends in restaurant menu development and marketing is the appearance of seasonal menus. These range from monthly to quarterly offerings of limited seasonal items. Figure 10-1 is from the seasonal summer menu for Ground Round restaurants featuring grilled items and sandwiches. Food photography provides illustration. The seasonal menu from the full-service dining room, Mary Elaine's, at the Phoenician Hotel in Scottsdale, Arizona, offers salads, appetizers, and entrees priced a la carte along with a fixed price (table d'hotel) "summer tasting menu" shown in Figure 10-2. Il Bistro Trattoria Il Panino in Boston, Massachusetts prints the weekly menu as in Figure 10-3 to reflect seasonal menu changes.

Mary Elaine's
Summer Tasting Menu

Grilled Australian King Prawns
Chilled Summer Vegetables with Pesto Sauce
Iron Horse, Fumé Blanc, Green Valley

Salad of Duck Confit with Apples and Fried Grapes
Pomegranate Syrup
Pine Ridge, Chenin Blanc, Napa

Seared Salmon with Confit Tomatoes
Fried Salads and Black Olives
Sanford Pinot Noir, Santa Barbara

Mesquite-Grilled Squab with Pancetta and Sage
Grilled New Potatoes and Vinegar Sauce
Kendall-Jackson Vineyards, Syrah, Grand Reserve

Tutti Fruiti, "Our Way"

72.00
Suggested wines are 3 oz. each and an additional $28.00.

6/95

Salads & Appetizers		Entrées	
❍ Roasted Sea Scallops with Braised Oxtail Quinoa Pickled Beets and Pistachio Oil	11.50		
		Filet of John Dory with Basil and Black Truffles "Fork Mashed" New Potatoes and Braised Celery	31.50
Veal Sweetbread "Pot Stickers" with Sweet and Sour Red Cabbage and Chanterelles Sherry Vinegar Sauce with Pine Nuts and Sultanas	10.00		
		Tenderloin of Veal with Roasted Garlic-Rosemary Jus Tartelette of White Bean Purée and Glazed Spring Vegetables	34.00
Risotto of Cave Creek Escargots and Mushrooms Glazed Asparagus and Aged Parmegiano Cheese	10.50		
		❍ Crispy Skin Salmon with Gazpacho Vegetable Napoleon Roasted-Tomatillo Sauce with Cilantro and Lime	29.00
❍ Chilled Dungeness Crab with Toasted Cous Cous Shaved Salad of Crisp Vegetables and Lobster-Horseradish Aioli	9.00		
		Roasted-Muscovy Duck Breast with "Dolce-Forte" Sauce and Seasonal Mushrooms Strudel of Vanilla-Spiced Peaches, Caramelized Shallots and Duck Confit	31.50
❍ Gratin of Herb and Spinach Cannelloni Sauté of Artichokes and Wilted Greens	9.00		
		Mesquite-Grilled Tenderloin of Beef with Osso Bucco Cannelloni Red Wine-Shallot Syrup and Pickled Leeks	31.50
Applewood-Smoked Salmon and Potato Crisps with Spicy Tuna Tartare Wasabi-Chile, Beluga Caviar and Celery Crème Fraîche	11.00		
		❍ Roasted Sea Bass with Caramelized Red Onion Ravioli Bouillabaisse Essence, Garlic Croûton and Whole Roasted Young Bok Choy	29.00
Cream of Lobster Soup with Morel and Porcini Mushroom Infusion Lobster and Wild Mushroom Ragoût	7.50		
		Garlic and Herb-Crusted Rack of Lamb with Tapenade Smoked Peppers and Chiles with "Socca Niçoise"	34.00
❍ Tian of Parmegiano and Grilled Vegetables with Tomato Compôte Fresh Buffalo Mozzarella with Roasted Eggplant and Pesto	9.00		
		❍ Daurade Royale with Garden Peas and Coconut Basmati Rice Sweet Lobster Curry Sauce with Fried Ginger and Mango	30.00
Salad of Braised Artichoke and Goat Cheese Gratin with Chilled Ratatouille Spiced Sumac Vinaigrette and Artichoke Coulis	9.00		
Warm Salad of Lobster and Shellfish New Potatoes and Green Beans in Citrus-Parsley Oil	14.50		
Chilled Maine Lobster with Avocado, Fava Beans and Hearts of Romaine Pink Grapefruit and Passion Fruit-Basil Vinaigrette	14.50	❍ *Choices – Superb cuisine created by Chef Alessandro Stratta with your well-being in mind.* *As a courtesy to other diners, please refrain from using cellular telephones and smoking cigars or pipes in the restaurant.*	

♦ **FIGURE 10-2.** A table d'hote tasting menu is teamed with the seasonal menu from Mary Elaine's restaurant at the Phoenician Hotel in Scottsdale, Arizona. Reprinted with permission of the Phoenician Hotel, Scottsdale, Arizona.

♦ **FIGURE 10-3.** A weekly menu reflects seasonal product availability from Il Bistro Trattoria Il Panino in Boston, Massachusetts.

DESSERT MENUS

An often overlooked area of the menu is the dessert section. Dessert sales can significantly increase the check average and corresponding revenues. Successful marketing programs to increase dessert sales offer a dessert menu and often a dessert tray with sample items for servers to introduce to customers. Dessert menus are often designed as table tent or stand-up pieces. The three menus featured in this section all act as table tents, placed on the table prior to customers being seated. Having an introduction to the dessert menu before and during the meal encourages menu sales. Additional exposure from a dessert tray or cart being passed continuously through the dining area increases dessert sales even more. Figure 10-4 is a three-panel fold featuring a "Mile High Homemade Pie" selection that the restaurant is well known for. Whole pie sales are also encouraged by the note in the upper left

◆ **FIGURE 10-4.** Increased dessert sales focus on pie in this menu from Chelo's in Providence, Rhode Island. Reprinted with permission of Ad Art Litho.

♦ **FIGURE 10-5.** This three-panel table tent offers a limited but varied selection of dessert items. Reprinted with permission of Ad Art Litho.

corner of the menu. Figure 10-5 uses appetizing illustrations of the feature items on each of the three panels to encourage sales. Figure 10-6 is a two-panel horizontal fold menu with the illustration of a banana split on the cover to identify the "Ice Cream Shop" theme of the menu. Colorful accent copy highlights the restaurant decor and theme.

♦ **FIGURE 10-6.** Ice cream based desserts are the theme of this sandwich and dessert menu from the Dolphin Hotel in Orlando, Florida. Reprinted with permission of Walt Disney World.

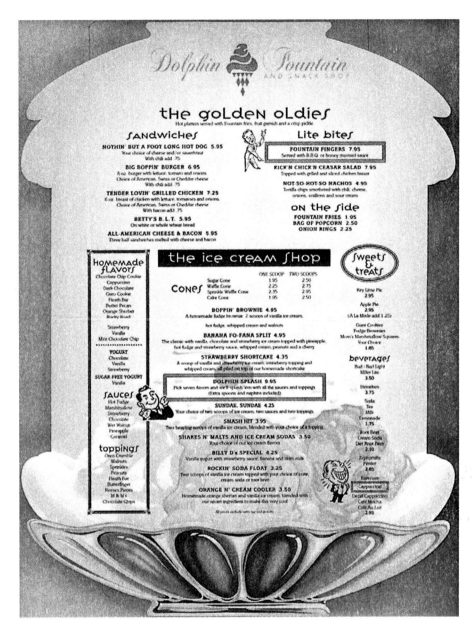

♦ **FIGURE 10-6.** *(Continued)*

BEVERAGE MENUS

Beverage menus, like dessert menus, can significantly increase check averages by creating interest in targeted beverage items such as frozen drinks, wine, cocktails, beer, and coffees. The cover for the wine menu in Figure 10-7 identifies wine labels featured on the menu. Hard Rock Cafe promotes frozen drinks made with top shelf liquors in Figure 10-8. By identifying specific liquor brands, management presells

♦ **FIGURE 10-7.** The cover of this center-fold menu from Sam's Cafe in Scottsdale, Arizona features wine bottle labels from the menu selections. Reprinted with permission of Sam's Cafe, Scottsdale, Arizona.

the customer on the higher priced drink and simultaneously has menu costs underwritten by the participating liquor companies. As this is an expensive multi-page menu with a spiral binder, menu production costs will be significant. Note the additional marketing of souvenir glasses at the bottom of both pages. Houlihan's specialty beverage menu promotes alcohol-based after-dinner coffees in Figure 10-9. The Iron Hill Brewery posts the daily microbrews on a central blackboard and

♦ **FIGURE 10-8.** Hard Rock Cafe uses a spiral bound, multi-page format for colorfully illustrated beverage menu. Reprinted with permission. Hard Rock Cafe is a registered trademark of Hard Rock Cafe Licensing Corporation.

♦ **FIGURE 10-9.** Houlihan's restaurant in Boston, Massachusetts promotes specialty alcohol-based hot coffee drinks using suggestive selling techniques.

IROn HILL BREWERY & RESTAURAnT

BEER TASTES BEST WHEn IT'S MADE FRESH, nOT HAVInG SPEnT WEEKS In A WAREHOUSE OR On A SHELF.
THAT'S WHY OUR BEERS ARE MADE RIGHT HERE, On A DAILY BASIS. WE USE OnLY THE FInEST InGREDIEnTS:
DOMESTIC AnD IMPORTED MALTED BARLEY AnD HOPS, YEAST AnD WATER. AnD BECAUSE WE USE nO
PRESERVATIVES, YOU ARE GUARAnTEED A FRESH AnD nATURAL GLASS OF BEER EVERY TIME.

1. **MILLInG:** Grains are selected depending on the beer style. The grains are milled into the grist hopper [A] to allow us to extract essential ingredients.

2. **MASHInG:** The grains are moved from the mill room through a feed auger [B] to the mash tun [C]. Hot water is added to form the mash. This process converts the grain starch to sugars.

3. **LAUTERInG:** Hot water is passed over the grains to remove all sugars. The solution, or wort, is extracted from the mash tun and sent to the kettle [D].

4. **BOILInG:** The wort is brought to a full rolling boil and hops are added for bitterness and aroma—giving each beer its unique profile.

5. **COOLInG:** The hot wort is passed through a heat exchanger [E] where it is cooled to a temperature appropriate for the yeast to ferment.

6. **FERMEnTATIOn:** The wort is transferred to a fermentation tank [F] where yeast is added and it is allowed to ferment, converting the sugars to alcohol and CO2. Depending on the type of beer, this process takes 14 to 30 days.

7. **FILTERInG:** After proper conditioning, the beer is sent through a filter [G] to remove all traces of the yeast before it is transferred to the serving tanks [H].

8. **SERVInG:** Finally, the beer carbonation level is adjusted and it is ready to be sent directly to our taps for consumption. From start to finish our beers travel less than 65 feet. There's nothing fresher!

nEWARK'S FIRST. DELAWARE'S FInEST.
147 EAST MAIn STREET nEWARK, DELAWARE 19711 302 266.9000

◆ **FIGURE 10-10.** An illustrated placemat introduces customers to the microbrewery concept at the Iron Hill Brewery and Restaurant in Newark, Delaware. Reprinted with permission of Iron Hill Brewery and Restaurant, Newark, Delaware.

uses the placemat in Figure 10-10 to create customer interest by providing an explanation of the brewing process. The mulit-page Italian wine menu in Figure 10-11 regionalizes the map of Italy to identify locations of the winery of origin. The sports bar menu in Figure 10-12a and b is inserted into a plexiglass stand for use as a table piece, as is the gourmet coffee menu.

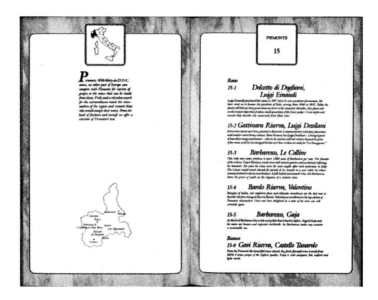

◆ **FIGURE 10-11.** A multi-page wine menu identifies the many regions of Italy from Ristorante Lucia, Boston, Massachusetts. Reprinted with permission of

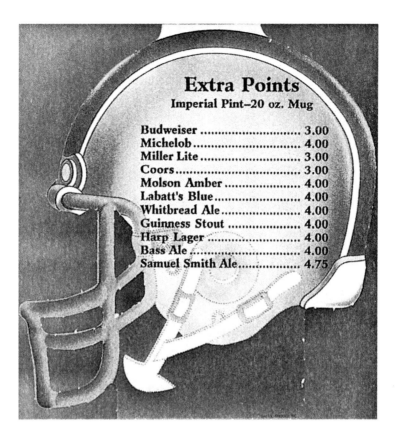

Extra Points

Imperial Pint—20 oz. Mug

Budweiser 3.00
Michelob 4.00
Miller Lite 3.00
Coors 3.00
Molson Amber 4.00
Labatt's Blue 4.00
Whitbread Ale 4.00
Guinness Stout 4.00
Harp Lager 4.00
Bass Ale 4.00
Samuel Smith Ale...................... 4.75

Gourmet Coffees

Served Plain 2.50
With Imported Syrup 3.50
With Liqueur 4.95

Espresso
Cappuccino
French Roast
Hazelnut
Mocha Java
Irish Cream
French Vanilla
Chocolate Almond
Chocolate Raspberry

◆ **FIGURE 10-12.** Table tents promote sports bar and gourmet coffee menus. Reprinted with permission of Gastro-Gnomes, West Hartford, Connecticut.

237

CHILDREN'S MENUS

Children have become important customers in the restaurant industry. Fast-food chains cater directly to the influence of their buying power. Family-style and full-service restaurants, understanding the importance of providing an enjoyable dining experience for parents, have developed colorful, game-oriented menus. Interestingly, most children's menus include the same menu items that have proved successful in other restuarants. Figure 10-13 is a placemat incorporating the children's menu and activities for a range of ages and abilities. Figure 10-14 from Ground Round restaurants is an exciting food photography menu that provides a pencil sketch activity on the reverse side. Chili's provides a seasonal children's menu and activity book shown in Figure 10-15, using chili-related figures and copy to tie in the restaurant theme. The children's menu from the Bubble Room in Naples, Florida uses imaginative text and illustrations on their children's menu.

♦ **FIGURE 10-13.** A children's menu offering a wide variety of activities. Reprinted with permission of Sam's Cafe, Scottsdale, Arizona.

◆ **FIGURE 10-14.** Pencil sketch is the activity theme of this children's menu from The Ground Round restaurants. Reprinted with permission of Ground Round Restaurants.

◆ **FIGURE 10-15.** Chili's provides a multi-page book format as a children's menu.

◆ **FIGURE 10-16.** The Bubble Room in Naples, Florida provides creative activities with amusing illustrations. Reprinted with permission of The Bubble Room, Naples, Florida.

CHAPTER 11

◆

The
Changing
Customer

PROMOTIONS

Promotion is the use of advertising or publicity to gain public acceptance of a product and has become a large part of the marketing effort. Foodservice promotions can be game cards with food and merchandising prizes, redeemable coupons for food, children's giveaway items, or theme menus and festivals combined with entertainment and associated beverages.

Interesting the customer is important, but realizing value from dollars spent is even more so. Promotions cost money to develop, advertise, and produce. Calculating the return on the invested promotion dollar cannot be done by looking at the immediate volume sales gains made during the promotion period. Promotion costs, including advertising, should be subtracted from volume sales gains. Long-term gain is measured over a prescribed period to determine new customer return rates and real volume interest. If, after a promotional period, sales figures return to normal and show no further substantial gains, then the promotion was not effective from either the customer's or management's point of view.

Thorough market research of current and future customers as well as the competition will help in developing valuable promotions that increase both customer counts and profits. Before promotional campaigns are developed, there are three questions to be considered:

1. What needs promoting?
2. Can alternatives to food and beverages be promoted (such as service and atmosphere)?
3. Does the promotion have to include food discounting or coupons?

An important aspect of menu promotion is developing specialty menus and tabletop pieces to attract customer attention, such as those featured in the discussion of dessert and beverage menus in Chapter 10.

ADVERTISING

Advertising lets the public know about products and/or services offered. Whether or not potential customers receive your information depends on how well the advertising was planned. Planning for advertising is often done with professionals, either in a media-sales or consulting capacity, who understand all of the aspects and ramifications involved. Any dollar spent should be budgeted for in relation to its probable return in sales. Many chain and independent restaurants budget advertising spending against projected or actual earnings, for example, allocating 2 percent of every sales dollar for advertising. As with any cost, advertising dollars represent a calculated percentage of every dollar spent. Based on monthly sales receipts of $25,000, the 2 percent advestising budget for the month would be $500. These dollars can be spent on a number of media forms, including newspaper or radio advertising, flyer and mailer printing and distribution, community development and promotions.

The potential effectiveness of each of these methods of advertising to reach the target market should be carefully considered. Planning an effective advertising campaign requires management to be familiar with the target market, the number of probable customers, and the overall geographical area that a media form can reach, as well as the total cost of using a particular media form. In addition, the period of time in which the medium will be effective should also be understood. For advertising dollars to be spent effectively, each of these points should be researched and thoroughly understood. Understanding the mechanics, terminology, and capability of the medium is necessary for successful marketing.

ANALYZING MENU EFFECTIVENESS

An analysis of menu performance is generally based on the final food and beverage cost percentages in a given period. If the percentages match the forecasted figures and customers are satisfied, it is often assumed that the menu program is doing well. Menu effectiveness, however, should not be assumed on the basis of such incomplete analysis. In addition to examining the end-of-period financial figures, the menu planner should review the following questions.

1. What percentage of the sales dollar does cost of food actually represent?
2. Are the forecasted percentages being reached?
3. Does the profit margin reflect proper management of food costs?
4. Are customer counts up?
5. Has the average check per service increased?
6. What is the sales mix of high-profit items to low-profit items?
7. What are the current customer counts and average check figures, as compared to the last period calculated?
8. What has the customer response been to the menu?

One negative answer to any of these questions does not necessarily mean that the menu program is performing poorly. Each question may have a number of considerations that should be taken into account. Finding more than one negative answer, however, does indicate that some of the foodservice's problems lie in the menu. Identifying the source of the problem is not always easy. Often, the answer is hard to pinpoint and considerable time may need to be spent uncovering information.

Analyze the negative answers. Where does the problem begin and what could be the most probable cause? For example, if food cost is too high, any of the following questions could reveal the source of the problem.

Is the problem in the kitchen? If so, is it caused by:

♦ overportioning?
♦ overproduction and waste?
♦ employee theft?

Is the problem in purchasing? Is it a result of:

- poor quality food?
- production problems?
- spoilage in storage?
- poor purchasing procedures for ordering and receiving?
- employee theft?

Is the problem in the dining room? Can it be attributed to:

- selling techniques—pushing low-profit, high-cost items?
- customer dissatisfaction, revealed by excessive amounts of wasted food being returned to the kitchen?
- employee theft?

These questions address only some of the possible causes of one menu problem—a food cost percentage that is too high. Similarly, while these types of problems are common to most foodservice organizations, each operation has individual characteristics that affect the success or failure of the menu. A systematic, objective analysis of the problem is essential.

Although a discussion of security issues is not within the scope of this book, it is helpful to remember that the foodservice industry suffers high losses because of employee theft, which is, unfortunately, a problem common to most foodservice operations. Computer software programs can help to identify this and other problem areas and provide the controls needed to alleviate or minimize them.

Menu analysis should be done as impartially as possible. Although it is difficult to criticize the results when so much time and effort have been spent on developing the menu, the menu selection, and the pricing structure, objective criticism is essential.

The effectiveness of a menu in today's highly competitive market is the barometer of a restaurant's success or failure and of management's ability to make each sales dollar as profitable as possible.

◆ GLOSSARY

abstract The costs of all foods such as table salt, pepper, sugar, etc., that are not part of any standard recipe card. It is generally calculated at 2 percent of the recipe's total portion costs.

accent copy In a menu, the use of creative names for dishes or the headings of courses to highlight or draw attention to these sections. Accent copy is one of the three kinds of written text, or copy, used in menus.

a la carte One of the four menu pricing formats. In an a la carte menu, each dish is priced individually and all items are grouped into courses, which are listed in the menu according to the order of service.

American service A style of table service in which each meal or dish is placed on a separate plate in the kitchen and then brought directly to the customer.

antipasto Italian appetizers, usually cold meats, cheeses, olives, peppers, and relishes.

assimilation The merging of diverse cultural elements.

average check The average total amount that each customer spends at a given meal service, such as lunch or dinner.

back of the house The area of the restaurant outside of the guest contact area, such as the kitchen, storeroom, back office, etc.

basic menu design The shape and size of a physical menu as well as the number and types of menu pages and panels used.

brightness The amount of reflection or glare that is given off by the finished surface of a paper stock.

buffet service A kind of table service in which all or part of a meal is set out on a table and customers help themselves to the food.

caliper measurement The bulk or thickness of a paper stock.

central purchasing list An inventory of all items that are used by the foodservice organization. From the central purchasing list, a shopping list can be made of items that must be purchased because supplies are low.

centralized feeding system A feeding system in which food is either fully or partially prepared in one central kitchen and distributed to auxiliary feeding stations or dining areas.

C factor Represents the cost of ingredients in a recipe that are valued at a penny or less.

clip art Uncopyrighted camera ready art work that can be cut out for use in advertisements, menus, and other printed materials.

coated paper Paper stock that has been treated with a layer of chemicals to form a smooth, reflective surface.

commercial foodservice organization　A foodservice outlet that is established for the purpose of making a profit.

commis de rang　In classical French table service, the assistant or bus person of the *chef de rang*.

Continental breakfast　A limited breakfast consisting of juice or fruit, a breakfast bread, and coffee or tea.

cooking load　The total amount of food that can be cooked by a foodservice operation at a given time.

cost per unit　The current wholesale price of a unit of an ingredient as purchased.

cuisine　The style or manner in which food is prepared.

decentralized feeding station　A foodservice system in which food is prepared for separate or individual dining areas in kitchens attached to the respective dining facility.

descriptive copy　One of the three kinds of written text, or copy, used in menus, descriptive copy is the written account of an individual dish, describing the item's ingredients or style of preparation.

display type　A typeface meant to catch the eye, such as those used for menu titles and headings.

embossed paper　Paper stock with a raised or textured surface.

en croute　A French term meaning *wrapped in pastry*.

entremet　French term for the dessert course.

expendable income　The segment of private income that remains after the basic needs of living have been paid for; the income that would be available for entertainment, dining out, and other such expenses.

extension　The cost of the amount of an item used in a particular recipe (total extension = overall food cost per recipe).

family-style service　A style of table service in which platters or serving dishes of food are placed at the table and customers help themselves to the food, passing the dishes to others at the table as needed.

fast-food operation　A foodservice operation, usually self-service, in which foods that require minimum preparation time are served.

feasibility study　An analysis of a proposed business operation to determine whether it will be practicable and profitable.

finish of paper　The smoothness or texture of processed paper stock.

fixed cost　A business cost that does not fluctuate.

food-cost percentage　The percentage of the total revenue or each sales dollar that represents the cost of food.

French service　A style of table service in which all or part of the food is prepared or finished at the customer's table by the chef de rang.

front of the house　The restaurant operations that take place in the dining room and adjacent service area of a foodservice outlet, such as taking orders, serving food, and collecting payment for the meal.

full-service operation　A restaurant offering style of cuisine, table service, atmosphere, and decor conducive to formal dining, as opposed to a fast-food outlet or self-service cafeteria.

grade of paper   The name designation of a paper type, such as cover paper, text paper, or bond.

grain of paper   The direction in which fibers run in a sheet of paper, analogous to the grain of wood or veneer.

"grazing"   A contemporary term for choosing a number of items from the appetizer course to be consumed as the main meal.

gross profit   The total profit before taxes have been subtracted.

haute cuisine   The highest style of classic French cookery.

industrial foodservice organization   A foodservice outlet usually established or commissioned by a business to meet the dining needs of its employees by providing a cafeteria, vending machines, or similar services.

institutional foodservice organization   A foodservice outlet that has been established to provide service in a nonprofit setting, as for a hospital, public school, or prison.

italic type   The slanted version of a typeface.

lamination   The bonding of a separate sheet of plastic to a piece of paper.

layout   The arranged placement of all the graphic and mechanical elements of a physical menu.

letterspacing   The amount of space between each letter in a typeset word.

lightening   The amount of white space that surrounds a letter or line of type; adding space to highlight or draw attention to a section of the menu.

losers and leaders   The food items that are at the bottom and the top of the sales-mix list. *Losers* are the items that sell most poorly; *leaders* are the best-selling dishes.

market survey   An evaluation of the target market, community, and competition of a given geographical area, conducted to determine whether a foodservice outlet opening in that area would be economically feasible.

menu cycle   An organized schedule for presenting different preplanned menus in a repeated pattern over a number of days, used primarily by institutional foodservice operations that serve the same people every day for an extended amount of time.

menu repertory   An indexed listing of all the menu items that a foodservice operation can serve. Each item in the menu repertory has a standard cost card and recipe card so that the main menu can be changed or updated easily without altering standards of quality, quantity, or costing.

merchandising copy   One of the three types of written text, or copy, used in a menu; provides basic information about the restaurant, such as the restaurant's address, telephone number, hours of service, and credit cards accepted.

net profit   The amount of profit after taxes have been subtracted.

overhead   Operating expenses, such as rent, taxes, utilities, and insurance, not including food and labor costs.

oversell   Using more means than necessary to persuade a customer to buy a product. In foodservice, oversell could be caused by a too-lengthy description of an item on a menu or by a verbose sales pitch made by a server.

panel   An unfolded section of the cover of the physical menu. A two-panel menu has the form of an open book; a tri-panel menu resembles a page folded into thirds.

paper weight   The number of pounds that 500 sheets (a ream) of paper weighs in a basic size; for example 20-lb stock, 80-lb stock.

par stock list   An inventory of all items kept in supply, along with the quantity of each item that should always be available in stock.

pâtissier, ière   French term for pastry chef.

pension   A type of boardinghouse where room and board are provided for a fixed fee.

perceived value   The customer's perception of how much money an entree or other dish is worth.

pièce de résistance   French term for the main dish of a meal.

portion   The amount allocated as an individual part or share; a serving.

prime cost   The combined total costs of raw materials and direct labor costs involved in producing a product.

prix fixe   French for *fixed price;* in a table d'hôte menu, the entire meal is offered at one fixed price.

production sheet   An itemized list of each food selection to be served at a given meal, with an estimate of the number of portions that will be served.

promotion   The use of advertising or publicity to sell or make a product or service popular.

random menu cycle   A schedule for presenting preplanned menus systematically in which a complete day's menu is given a letter or number and placed in sequence with the other menus to be served. The schedule can be changed by rearranging the order of the letters or numbers randomly.

ream   The quantity in which paper is packaged and sold. One ream equals five hundred sheets of paper.

remove   A term used in seventeenth-century table service to indicate a first course item that would be removed and immediately replaced by another.

revenue   The total income generated by a business.

roman type   The usual, upright form of a typeface.

Russian service   A style of table service in which food is presented to the customer on a serving platter and then added to the plate at the table.

sales history   A foodservice operation's daily log of total revenue, customer counts, overall costs, and any special events that might affect business, such as the weather.

sales mix   A chart in which the daily sales of each food item are recorded.

semi à la carte   A menu pricing format in which the price of one item, such as an entreé, would include other items, such as a salad and starch.

semi table d'hôte   A menu pricing format in which most of the meal is offered at one price and one or two courses, such as the appetizer and dessert, are priced separately.

sorbet A light ice usually made with fruit and approximating sherbet.

specifications An exact description of an item to be purchased, listing the item's size, weight, measure, count, or other characteristics. Specifications ensure that the correct item is purchased.

split-menu cycle A schedule for presenting menus systematically, in which each item on the menu is given its own cycle, to be used in conjunction with all other items to form a meal.

standard cost card An itemized breakdown of a recipe by ingredient, with a cost assigned to each ingredient, however small. Cost cards also list the cost of the recipe per portion and per yield.

standard portion The size of measurement of each individual serving of a standard recipe.

standard recipe A set of instructions that has been systematically tested and proven to result in a food product of consistent quality and quantity, when produced under a given set of circumstances.

standard yield The total number of portions that will be produced by a standard recipe.

table d'hôte A menu pricing format in which an entire meal is offered at one price. Often, the restaurant menu will offer only one meal per service.

tapes bar A Spanish cuisine bar/restaurant, which traditionally serves small appetizer portions of menu items along with sherry and wines by the glass.

target market The section of the consumer market toward which sales efforts are directed.

text type Smaller-sized typefaces used generally for menu items and descriptive copy.

three-color-vision theory The theory that all colors are based on the three primary colors: yellow, red, and blue.

tip-ons Cards or notes that are attached to a menu by means of a metal or plastic clip or an adhesive board. Tip-ons are used to inform the customer of changes in the menu, daily specials, or other information.

transfer tape Sheets of type that adhere to a paper surface when pressure is applied. Also known as press type.

truth in menu The legal regulation stipulating that the item or dish that a restaurant serves be exactly what the menu or server claims it is.

turnover The rate at which seats in a restaurant are occupied by new customers.

typeface A particular style of type, such as Bodoni, Helvetica, and so on.

type weight Refers to the degree of thickness of a typeface (light, medium, heavy, bold, etc.)

typical-break menu cycle A schedule for presenting different preplanned menus in a pattern that repeats on staggered days of the week. A six-day typical-break menu cycle could be used for a five-day serving plan, for example, so that the first menu would be served on a Monday in the first week, then served on a Tuesday the second week, and so on.

typical menu cycle A schedule for presenting different preplanned menus in a fixed pattern, such as a menu for each day of the week. The pattern is repeated without change.

typography   The art of composing printed matter from type.

uniformity of design   In developing a physical menu, the practice of using design elements (such as color, graphics, illustrations, and typefaces) that may be different but are compatible with one another to create an impression of consistency and uniformity in the overall design.

value of a color   Refers to how light in tone a color is.

yield   The amount of product that can be obtained from either a recipe or a raw ingredient.

◆ BIBLIOGRAPHY

Apicius. *Cookery and Dining in Imperial Rome,* ed. Joseph Dommers Vehling. New York: Dover Publications, 1977.

Axler, Bruce H. *Foodservice: A Managerial Approach.* The National Institute for the Foodservice Industry, 1979.

Combes, Steven. *Restaurant French.* 2d ed. London: Barrie & Jenkins Ltd., 1975.

Crawford, Hollis W., and Milton C. McDowell. *Math Workshop for Foodservice/ Lodging.* New York: Van Nostrand Reinhold, 1980.

Garry, William. "Democracy a la Carte," *Bon Appetit* September 1995: 14.

Hale, William Harlan. *The Horizon Cookbook and Illustrated History of Eating and Drinking through the Ages.* New York: American Heritage Publishing, 1968.

Kasavana, Michael L., and Cahill, John J. *Managing Computers in the Hospitality Industry.* East Lansing, Michigan: Educational Institute of The American Hotel & Motel Association, 1997.

Kasavana, Michael L., and Smith, Donald Y. *Menu Engineering: A Practical Guide to Menu Analysis.* Lansing, Michigan Hospitality Publishers, 1990.

Kotschevar, Lendal H. *Management By Menu.* 3rd ed. Dubuque, Iowa: Kendall Hunt, 1993.

Kreck, Lothar A. *Menus: Analysis and Planning.* 2d ed. New York: Van Nostrand Reinhold, 1984.

Miller, Jack E. *Menu Pricing and Strategy,* 4th ed. New York: Van Nostrand Reinhold, 1996.

Montagne, Prosper. *The New Larousse Gastronomique.* New York: Crown Publishers, 1988.

Pocket Pal: A Graphic Arts Production Handbook, 12th ed. New York: International Paper Co., 1979.

Reid, Robert D. *Hospitality Marketing Management,* 2d ed. New York: Van Nostrand Reinhold, 1989.

Roberts, Laura. *Rhode Island Historical Society Information for Tour Guides: Dining at the John Brown House,* 1980.

Scanlon, Nancy A. *Catering Menu Management.* New York: John Wiley and Sons, 1992.

———. *Menu For Profit* Video Program. N. New Portland, Maine: Profit Enhancement Programs, 1991.

———. *Restaurant Management.* New York: Van Nostrand Reinhold, 1993.

Seaberg, Albin G. *Menu Design, Merchandising and Marketing,* 4th. ed. New York: Van Nostrand Reinhold, 1991.

◆ INDEX